# Animal Breeding and Genetics

# Animal Breeding and Genetics

*Editor*

**Carin Fisker**

scitus
academics

**Animal Breeding and Genetics**
Edited by **Carin Fisker**

ISBN: 978-1-68117-415-0
Library of Congress Control Number: 2016937426


www.scitusacademics.com
Box No. 4766, 616 Corporate Way,
Suite 2, Valley Cottage,
NY 10989



Printed in United States of America on Acid Free Paper

# Preface

From the very early days human beings depend on animals and animal products for food and other requirements. In dairy and poultry farms high yielding animals are reared. These high yielding animals are produced by hybridization experiments. Previously the animals were developed basing on unscientific methods. Before the discovery of principles of heredity human beings have selected the animals with required characters and learned to develop the plants having the selected characters. This phenomenon is called Artificial selection. However, an increased knowledge of biology, especially genetics, has helped in improving the quality of animals and animal products as per the human requirements. The revolution in genetic mapping technology and the advent of whole genome sequences have turned quantitative genetics into one of the fastest growing areas of biology. The animal breeding and genetics provide new scientific discoveries to age-old livestock production problems to help producers and consumers. Animal breeding addresses the evaluation of the genetic value of livestock. Selecting for breeding animals with superior EBV in growth rate, egg, meat, milk, or wool production, or with other desirable traits has revolutionized livestock production throughout the world. The scientific theory of animal breeding incorporates population genetics, quantitative genetics, statistics, and recently molecular genomics. The book animal breeding and genetics encompasses topics such as genetic variability, genetic testing, and animal breeding focuses on various aspects of animal heredity, or the passing of traits from one generation to the next. It is of valuable tool for students, researchers, professors and a variety of employers, including government agencies, zoos, and food producers.

# Table of Contents

# CHAPTER 1

# A Novel Discipline in Embryology — Animal Embryo Breeding

**Bin Wu[1, 2], Linsen Zan[3], Fusheng Quan[4] and Hai Wang[1, 2]**

[1] Arizona Center for Reproductive Endocrinology and Infertility, Tucson, Arizona, USA

[2] Yunnan Jiuzhou Hospital, Kunming, Yunnan, China

[3] College of Animal Science and Technology, National Beef Cattle Improvement Center, China

[4] College of Veterinary Medicine, Northwest A&F University, Yangling, Shaanxi, China

## ABSTRACT

The modern animal biotechnologies, such as animal cloning, transgenesis, sex determination, stem cells, designing new livestock, must be performed on animal gametes including sperm and oocytes, and embryos based on embryology theory. Currently, some key biotechnologies in embryology have become the most powerful tool for animal scientists and breeders to improve genetic construction of animal herds. Here, authors put forward a new concept of Animal Embryo Breeding Science to describe this discipline formation, development, and application in animal genetic improvement and breeding. The relationship of embryo breeding with other disciplines has been profiled. Thus, animal scientists and breeders can easily understand and apply embryo breeding theory and related key techniques to accelerate animal improvement speed, to modify genetic construction of animal population, and to design and create new animal individual or breed.

## KEYWORDS

Discipline, embryo breeding, biotechnology, livestock

## 1. INTRODUCTION

Animal breeding sciences concern the management and care of farm animals by humans for profit. Not only does it refer to the practice of selectively breeding and raising livestock to promote desirable traits in animals for utility, sport, pleasure, or research [1], but also it refers to the efficient exploitation of a species in agriculture advantageous to humans. The genetic improvement of livestock depends on defining breeding objectives and accurately identifying the right animals to be used for future breeding. Traditional breeding programs involve 1) the design of animal breeding goals including improvement traits, such as milk, wool, growth, carcass and fertility, females vs. males, progeny test and nucleus vs. commercial animal population; 2) application techniques, such as artificial insemination and embryo transfer, are used as methods not only to guarantee that females breed regularly but also to help improve herd genetics; 3) based on quantitative genetics theory, estimation of breeding value by phenotype, pedigree, BLUP (best linear unbiased prediction) method, and genetic markers; 4) selection and culling of individuals based on genetic evaluation, balancing rate of change, and inbreeding; and 5) determining mating system. This is a long-term process for livestock genetic improvement.

As modern biotechnology develops, some new techniques can be applied to animal breeding programs 1) to accelerate genetic progression by shortening generation interval and increasing female reproduction; 2) to add new genetic trait to animal body by transgenic technology or to remove bad traits from animal body by gene knockout method [2]; and 3) to create new animal individual or breed by modern biotechnologies including nuclear transfer, cloning, and genetic modification. These new technologies will make it easier to manipulate animal genomes, but food products from genetically engineered animals face a long road to market. Examples of biotechnology applications of particular interest to the department include cell culture, genomics, molecular-marker-assisted breeding, cloning, bioprocessing, and diagnostic testing, as well as gene technology (genetic modification). Genetic modification deliberates change of an organism's genetic material by moving, introducing, or eliminating specific genes, such as taking a single gene from an animal cell and inserting it into another animal cell to give the second cell a desired characteristic. The terms "gene technology," "genetic engineering" and "genetic manipulation," "genetic enhancement," "gene splicing," "transgenics," or the use of "recombinant DNA" are used to describe genetic modification processes. Genes can be found in and moved between different plants, animals or microorganisms such as viruses or bacteria, for example, transferring worm fat-1 gene to pig to produce more omega-3 fat acid in pork meat [3]. Genes can also be changed within a specific plant or animal individual. For instance, "knocking out" an undesirable characteristic gene such as susceptibility to a particular disease can be beneficial to the plant or animal life.

In mammals, the realization of these goals must depend upon *in vitro* manipulation of animal oocytes and embryos. Thus, embryology has become a core of these biotechnologies (Figure 1). Currently, embryo biotechnology, which most people call **embryo bioengineering,** has gradually become the most powerful tool for animal scientists and breeders to improve genetic construction of their animal herds or populations. Embryo transfer in cattle has recently gained considerable popularity with seedstock dairy and beef producers. Many kinds of species have been cloned and some transgenic animals have been produced. Thus, embryology has become a core of modern biotechnologies in animal genetic modification and breeding. Combining the new advances in modern biotechnology with future application, authors put forward the new concept of **Animal Embryo Breeding Science** to describe embryology development and application in animal genetic improvement and breeding.

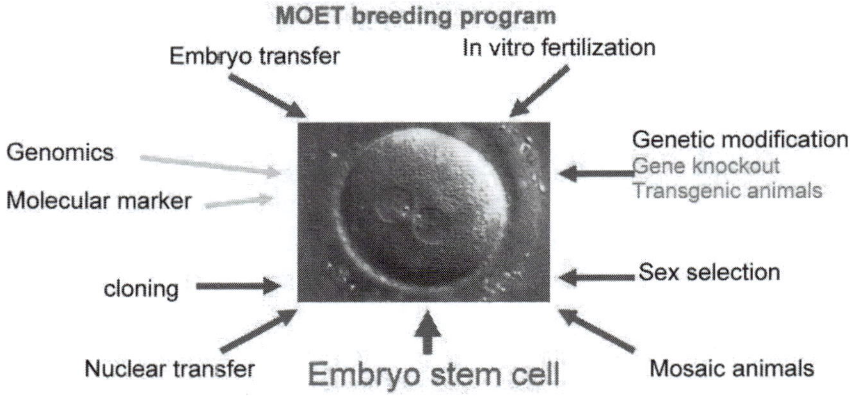


**Figure 1.** Embryology has become a core of modern biotechnologies in animal genetic modification and breeding. Any new developed biotechniques such as nuclear transplantation, cloning, and transgenesis, finally must be performed on animal oocytes or embryos. MOET represents multiple ovulation and embryo transfer.

## 2. CONCEPT OF ANIMAL EMBRYO BREEDING

Breeding is the reproductive process, which is producing of elite offspring in animals or plants. Animal breeding programs involve the selection or culling of parents (such as bull and cow) and then determination of mating system. They must be female and male sex combination. However, **Animal Embryo Breeding** is an asexual reproduction of specific oocytes or embryos artificially by current developed biotechnology. The Science of Animal Embryo Breeding is to study how to use the embryo manipulation technologies to improve, create, and clone new animal individual or breed. Current developed techniques include nuclear transfer, cytoplasmic transfer or replacement, in vitro fertilization (IVF), sperm cytoplasmic injection (ICSI), parthenogenesis and androgenesis, embryo cloning, sex selection, transgenesis, gene knock out, stem cells and somatic cell cloning, etc.

Although embryo breeding is a branch of traditional animal breeding discipline, the science of animal breeding is concerned with the application of the principles of population genetics and qualitative genetics to the improvement of domestic animals. However, Animal Embryo Breeding is concerned with application of the developed embryo biotechnologies to new animal individual creation, genetic cloning and preservation of animal breeds. The research main body of this discipline focuses on sperm, oocyte and embryo. After the desired animal type (genetic improved goal) has been designed, by means of a serial micromanipulation on oocyte or embryo, such as nuclear transfer, foreign DNA microinjection to egg pronucleus and stem cell technique, a modified improved embryo may be produced in vitro and then transferred into animal uterine cavity so that a new animal individual could be created. In the last couple of decades, many kinds of animals including transgenic pigs, cattle, sheep, and goat, have been produced [4].

# 3. THE RELATIONSHIP OF ANIMAL EMBRYO BREEDING SCIENCE WITH OTHER DISCIPLINES

As a new developing subject, Animal Embryo Breeding Science mainly depends upon modern biotechnology development, especially molecular biology, genetics, and reproductive biology with embryology. However, it also has a close association with other subjects such as reproductive biology and embryology, animal genetics and breeding (Figure 2).

The goal of animal breeding program can be realized by the current embryo breeding technology. Using molecular biological technique, a specific gene type for the desired animal may be designed. The new developed biotechnologies to attempt to modify animal genetic traits must be conducted on animal oocyte and embryo. The embryo in vitro production and animal individual birth must depend upon animal reproductive technology. Embryology may supply a good condition to produce many high-quality embryos. Thus, the Embryo Breeding is a core subject which combines molecular biology/genetics with animal genetics and breeding as well as reproductive biology and embryology.

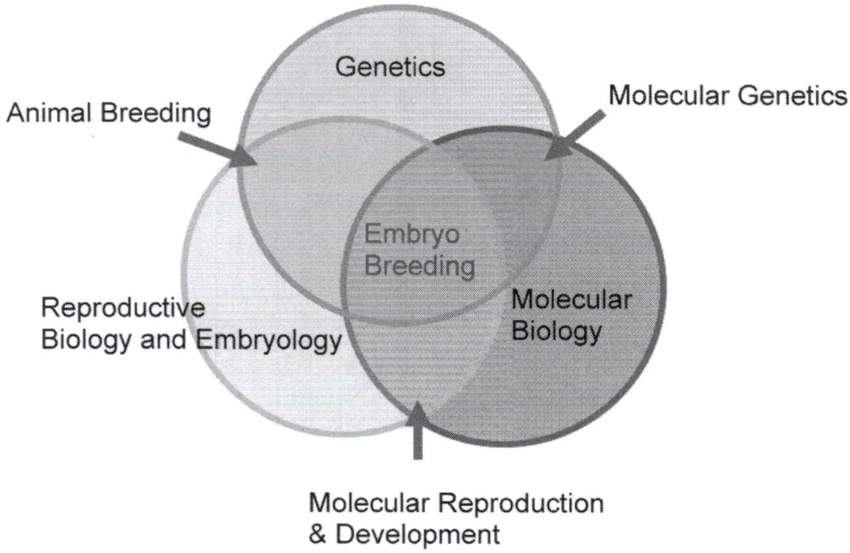


**Figure 2.** The designed relationship of Animal Embryo Breeding with other disciplines. The Embryo breeding is a core subject which combines molecular biology/genetics with animal genetics and breeding as well as reproductive biology and embryology.

# 4. MAJOR RESEARCH SCOPE AND CONTENT OF ANIMAL EMBRYO BREEDING

Animal Embryo Breeding Science is based on the current developed embryo biotechnology. The core of current embryo biotechnology is oocyte in vitro fertilization (IVF). As human IVF technique rapidly develops in infertility treatment, not only animal IVF has offered a very valuable tool to study mammalian fertilization and early embryo development, but also its commercial applications have being increased. Based on IVF research, some new developed embryo technologies consisting of nuclear transfer, transgenesis, cloning, and stem cells, etc., can be used to create new animal individual or population, and accelerate genetic progression of animal population during the period from early oocyte stage (oogenesis) to preimplantation embryo stage (Figure 3).

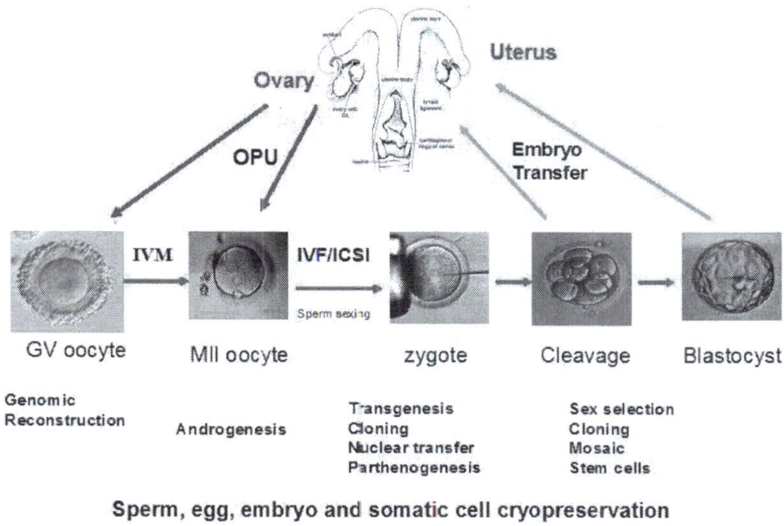


**Figure 3.** Schematic representation of main embryo biotechnologies which can impact on the genetic improvement programs on animal embryo breeding.

Based on this schematic picture, we may focus on several fields for Animal Embryo Breeding research. In the early stage of oogenesis and oocyte maturation, some key techniques such as genomic reconstruction, nuclear transfer, androgenesis and parthenogenesis, cytoplasm replacement, etc., may be used to change animal genetic construction [5]. At the fertilization stage, the sexing sperm may be used to produce specific-sex (female or male) animal population to achieve better economic results [6]. Using intracytoplasmic sperm injection (ICSI) technique may make an elite performance bull with a very few sperm produce a lot of offspring. At the pronuclear stage, the foreign DNA may be injected to zygote to produce transgenic animals. In the preimplantation cleavage and blastocyst stage, preimplantation genetic diagnosis (PGD) or preimplantation genetic screening (PGS), embryo cloning, mosaic animal and embryo stem cell techniques may be used to produce various different types of animals. Also, at any stage, sperm, egg and embryo, as well as somatic cells may be cryopreserved for future use

[7]. Thus, we may profile the outline of Animal Embryo Breeding study as shown in Table 1 (Table 1).

**Table 1.** Outline of Animal Embryo Breeding discipline

| | |
|---|---|
| **Early Gamete Manipulation** | |
| | *Artificial Insemination* |
| | Semen collection and its storage |
| | Sperm sexing |
| | Ovulation control |
| | Superovulation |
| | Ultrasound-guided oocyte retrieval (TVOR) or nonsurgical ovum pick up (OPU) |
| | *Oocyte and egg cryopreservation* |
| **Embryo Transfer** | |
| | Multiple ovulation (superovulation) |
| | Multiple ovulation with embryo transfer (MOET) |
| | Embryo splitting |
| | Embryo sexing |
| | Embryo transfer technique |
| ***In vitro embryo production (IVP) technology*** | |
| | *In vitro* maturation (IVM) of oocytes |
| | *In vitro* fertilization (IVF) of oocytes |
| | Intracytoplasm sperm injection (ICSI) |
| | Culture of *in vitro* fertilized embryos |
| | Preimplantation embryo diagnosis |
| **Embryo Cloning** | |
| | Embryo blastomere cloning |
| | Somatic cell nuclear Transfer (Dolly) |
| | Embryonic stem cell nuclear transfer |
| | Induced pluripotent stem cells (iPS) nuclear transfer |
| **Transgenic animals** | |
| | **Transfer gene construct** |
| | Inserting genes |
| | Knockout genes |
| | **Strategies for gene transfer** |
| | a) Directly inject a gene into egg pronucleus |
| | b) Sperm-mediated gene transfer |
| | c) Stem-cell-mediated gene transfer (transfection) |
| | d) Retrovirus and viruses vector for gene transfer |

| | |
|---|---|
| | e) Transfer of animal cells/embryo |
| | f) Targeted gene transfer |
| | g) Liposomes or spheroplasts as vector |
| | h) Other techniques such as electroporation, use of complexes, of DNA with polycations or lipids; a particle gun, DNA with polycations or lipids, etc. |
| | **Production of various transgenic animals** |
| | a) Cow or cattle |
| | b) Sheep and goat |
| | c) Fish |
| | d) Pig |
| | e) Other animals |
| | **Animal bioreactor and molecular farming** |
| | **Transgenic breeding strategies** |
| | **Transgenic effect and cost** |
| **Genomic reconstruction** | |
| | Germinal vesicle (GV) transfer |
| | Androgenesis |
| | Parthenogenesis |
| | Three-parent baby |
| | Ova-plasma transfer |
| **Preimplantation embryo diagnosis/screening (PGD/PGS)** | |
| | Fluorescent *in situ* hybridization (FISH) |
| | Polymerase chain reaction (PCR) |
| | Microarray |
| | Comparative Genomic Hybridization (CGH) |
| | Gene chips |
| **Mosaic animal creation** | |
| | Heteromorphosis |
| **Rare animal individual or breed preservation** | |
| | Sperm cryopreservation |
| | Egg cryopreservation |
| | Somatic cell cryopreservation |

# 5. RESEARCH CATEGORY OF ANIMAL EMBRYO BREEDING

As a new discipline, animal scientists and breeders can apply Animal Embryo Breeding Science theory to animal population to improve genetic traits, to add new benefit traits to animal body and to remove some harmful traits from animal body. Major research categories involve the following several aspects:

1. The objective of embryo breeding study is to create new animal individual or improve animal population. Based on the objective of the animal breeding program – what kind of animal traits you need in the breeding program – you may adopt an appropriate method of embryo biotechnique. For instance, if you want to add new genetic trait into animal body, you may use transgenic method to insert this gene into embryo. If you need to produce a complete same animal, clone method may be used as embryo cloning or nuclear transfer technique to clone this animal somatic cell.

2. The technique selection of embryo breeding: based on your breeding objective, a specific technique should be selected; for instance, in transgenic program, what gene and which method should be used to produce transgenic animals. In the animal cloning program to increase animal population homogeneity, various cloning methods should be evaluated for the best cloning technique, such as embryo cloning, stem cell, or somatic cell cloning.

3. Inserting embryo breeding into animal breeding program. In practice, embryo breeding is a trick to produce a specific animal. By the means of transgenic tactics, a given target gene vector may be constructed and transformed to chromosome in cell. Then, a given aim-gene embryo may be formed by nucleus transfer technique. By means of the genetic screening and diagnosis on cell levels, an expected embryo with a specific genotype embryo may be determined on embryonic level. Then, this expected embryo with a specific modified gene may be transferred into animal uterus to produce a specific animal. After individual level diagnosis, the ideal animal may be placed in animal population to expend its reproduction as traditional breeding program.

# 6. APPLICATION OF EMBRYO BREEDING IN ANIMAL IMPROVEMENT PROGRAM

- Genomic reconstruction by somatic cloning and parthenogenesis to produce specific animal population

When a bull or cow with elite production performance in beef cattle population is discovered, the breeding aim will be to accelerate this cow or bull reproduction to propagate a new breed of cattle. By normal breeding mating, this cow may lose half its inherent genes in its offspring. However, by the means of cloning techniques, many individuals of the same genotype can be theoretically produced. Thus, the accuracy of evaluation may be greatly increased. In spite of low cloning efficiency, many scientists are still interested in animal cloning techniques, which will eventually be used to clone very valuable animals, such as breeding stock, transgenic animals, and endangered species.

By the means of cell nucleus transfer technology, a new animal can be produced using androgenesis method [7]. Androgenesis is a male parthenogenesis in which only paternal chromosomes are kept in the embryo with the removal of the egg nucleus at the fertilization [8]. This is a reproductive pattern from two male parents. After an oocyte nucleus has been removed, a male diploid cell is transferred into this egg in which the oocyte cytoplasm will induce this diploid cell going through meiosis to become a haploid MII oocyte. After inducement, a male sperm is injected into this oocyte to produce a paternal embryo. Finally, this modified embryo will be transferred into receipt cow to produce a new individual bull with two male parents.

- Create new genetic variation in population by genomic modification during embryogenesis

The current animal breeding strategies are mainly based on the principle of selective breeding including the morphology of animal body, the application of quantitative genetics theory, the estimation of breeding value by phenotype, pedigree, BLUP (best linear unbiased prediction) method, and genetic markers. These methods mainly add genetic improvement by increasing the frequency of advantageous alleles of many loci, but actually very few of gene loci are identified. These techniques do not change gene movement from different species or genera due to reproductive barrier, while the new developed transgenic technique can remove the breeding barriers between different species or genera.

The most efficient method of transgenesis in mammals is the genetic manipulation of the pronuclear stage embryo [9]. By injecting foreign DNA into one of the two pronuclei of the zygote, the birth offspring may contain a functional foreign gene in the genome. In the last 20 years, many kinds of transgenic species have been produced for agriculture and medicine application [10]. For example, the transgenic technology in beef cattle industry may improve animals for faster growth, higher quality beef products, or disease resistance [11-13].

The transgenesis first starts with identification of the genes of interest. Current molecular biotechnology may help us to search for some interesting markers used as reference points for mapping relevant genes. These molecular markers can also be used for identification of the animals carrying the transgenes. Most of the quantitative genetic loci (QTL) are polygenic in nature but the manipulation of transgenesis is a single gene trait [14,15]. The technology holds promises in the future in moving polygenic QTL across the breeding barriers of animals. However, it is expected that molecular markers will serve as a potential tool to geneticists and breeders to evaluate the existing germplasm, and to manipulate it to create animals of desired traits [16].

- Shorten generation interval by embryo in vitro production

As the oocyte in vitro maturation (IVM) and in vitro fertilization (IVF) techniques rapidly develop, the ultrasound-guided oocyte retrieval (TVOR) or nonsurgical ovum pick up (OPU) technique can retrieve many oocytes repeatedly from a cow or a heifer. As many as 1000 oocytes have been collected from one female cattle in a year [17-19]. Thus, the embryo in vitro production (IVP) technology has been able to promote a cow to produce more than one hundred offspring in a year and greatly accelerate herd genetic improvement speed [20]. In order to improve ordinary cattle herd, slaughterhouse ovaries also may be used as in vitro embryo production. A lot of oocytes could be obtained from slaughter house cow ovaries. After maturation, these oocytes may be inseminated with elite bull semen for in vitro fertilization [21]. Although the detail genetic backgrounds of these slaughterhouse animals are not known, these embryos have a very high genetic merit from elite bulls. Using these embryos, an ordinary cow herd could obtain at least 50% genetic improvement.

The multiple ovulation and embryo transfer (MOET) was used initially to produce more embryos from genetic elite cows in shorter time periods. Currently, the MOET breeding schemes have widely established in many countries and their use accounts for about 80% of cattle embryos transferred commercially [22]. Currently, the application of transvaginal ultrasonically guided OPU technique may significantly improve MOET scheme efficiency because about 1000 oocytes may be collected and 300 embryos may be produced *in vitro* from a cow in a year at frequent intervals using IVF technology [19]. Also, oocytes may be collected from prepubertal heifers and cattle generation interval may be shorted for 2-3 years. The combination of MOET program with OPU/IVF technique is providing a more efficient way to produce more embryos from an individual donor donor than superovulation stimulation program [23]. Thus, OPU/IVF technique greatly increases MOET breeding scheme efficiency in milk and beef industry.

- Increased economy from animal population by sex selection

Animal sex selection may increase animal economical value for humans. Embryo breeding theory may provide several ways for animal sex selection, including sperm sex selection and preimplantation embryo sex selection. Sperm sex selection is to try to separate semen into X- or Y-bearing chromosome sperm by flow cytometry [24, 25]. Current sorted sperm has been successfully used in IVF for in vitro embryo production and artificial insemination in cattle [6, 26].

Another sexing pathway is to determine the sex of an embryo prior to transfer. Preimplantation genetic diagnosis (PGD) technique has become an efficient method for sex selection. Y-specific chromosome probe for polymerase chain reaction (PCR) and Fluorescent *In Situ* Hybridization (FISH) are two common methods in animal sex determination. On the ordinary farm, cattle embryos may be sexed by complete cell biopsy and PCR technique. Our clinic farm practice [7] showed that a few of trophoectoderm cells could be microbiopsied from blastocyst embryos by transzonal incision using a microsurgical blade. The mini-tube PCR was carried out for 30 minutes and the gel electrophoresis was run for 20 minutes. The sexing result could be obtained in 2 hours. These results clearly demonstrate that the microsurgical technique and subsequent PCR sex analysis allow the rapid commercial exchange of genetic resources on the basis of fresh or frozen sex-desired embryos in embryo transfer programs.

Fluorescent *in situ* hybridization (FISH) technique has also been used as embryo chromosome set (karyotype) diagnosis. A blastomere is removed from an embryo by micromanipulation, and then used to examine the embryo X/Y chromosomes by FISH. Recently, new developed technologies in PGD allow examining of all chromosomes and identifying certain genes or genetic mutations, such as the competitive genomic hybridization (CGH) and microarray analysis. More recently, novel developed Next Generation Sequencing (NGS) for preimplantation genetic screen (PGS) is now being offered clinically to provide comprehensive, accurate screening of all 24 chromosomes for selections of euploid embryos. PGS results generated are comparable to those achieved with the CGH technology, with improved accuracy, sensitivity, and resolution for more accurate detection of euploid embryos, aneuploidies, chromosome imbalances (translocations), and embryo mosaicism. NGS is a superior technology because it looks at close to 1.1 million data points on the genome compared to around 3,000 with CGH.

- Preservation breeding

Many animal breeders are interested in preserving bloodlines of animals, either of a rare breed, or of rare pedigrees within a breed. Therefore, Rahbek [27] put forward a preservation breeding concept to describe the purpose of preservation breeding, which is to protect genetic diversity within a species, and to preserve valuable genetic traits that may not be popular or in fashion in the present, but may be of great value in the future. In the animal embryo breeding program, two kinds of cells including reproductive cells and somatic cells may be cryopreserved in liquid nitrogen for future use. Reproductive cell cryopreservation is an important branch of embryo breeding science because it involves the preservation of gametes (sperm and oocytes), embryos, and reproductive tissues (ovarian and testicular tissues) for future use in the assisted reproductive technology. Practically, animal embryo breeding program may provide a sperm and embryo bank with the objective of avoiding genetic dilution and irreplaceable gene losses of the valuable "naturalized breeds" germplasm. It is much lower in cost than normal animal breeding, preserving rare native animal breed plan. At present, many countries have set up gene banks to store frozen embryos and semen of various animal species including native cattle, pig, and some endangered animals.

The development of embryo freezing technologies has revolutionized cattle breeding. Since then, advancements in cryobiology, cell biology, and domestic animal embryology have enabled the development of embryo preservation methodologies for our other domestic animal species, including sheep and goats. Currently, use of preserved embryos has become a routine breeding alternative for all domestic animal species. This freezing and storage methodology may provide for maternal germplasm, global genetic transport, increased selection pressure of herd genetics, and genetic resource rescue.

In the conventional breeding program, an outstanding bull may maintain normal mating for 5 years. However, if this bull semen is cryopreserved, it will extend the bull's breeding time. In embryo breeding program, when some elite bulls leave very few sperm, we may use intracytoplasmic sperm injection (ICSI) technique to inject a single sperm to an oocyte so that genetic merit embryos are obtained [28]. Also, sperm cell genome cloning technique may be used to produce many copies of a specific sperm [8]. The application of this technique to beef and dairy cattle industry has greatly increased merit bull spread in animal herd [29].

Like normal reproduction, somatic cell nuclear transfer (SCNT) starts with an egg or oocyte, but here the nucleus of the egg needs to be removed. Then the nucleus from a somatic (skin) cell is transferred into the enucleated egg which would be analogous to the sperm entering the oocyte. As this develops into a blastocyst, cells from the inner cell mass can be isolated and purified to serve as a source for pluripotent stem cells. In animal embryo breeding, somatic cell is also an important genetic resource. Therefore, the somatic cells, such as skin, hair, and other cells from rare and endangered animals may be collected and cryopreserved so that they can be used in the future.

# 7. CONCLUSIONS

Currently, the following biotechnologies in embryology have been applied or will be applied in animal genetic improvement [9]: 1) Genomic reconstruction by somatic cloning and parthenogenesis can produce specific animal population; 2) new genetic variation in population can be created by genomic modification during embryogenesis, such as transgenic breeding strategies; 3) animal generation interval may be shortened by embryo in *in vitro* production; 4) economy efficiency from animal population may be significantly increased by embryo sex selection; and 5) a rare breed, or of rare pedigrees within a breed, may be efficiently preserved at low cost in liquid nitrogen. Thus, the development of modern biotechnology has brought into being the concept and theory of **Animal Embryo Breeding Science.** Understanding and applying its theory and technology will be helpful to animal scientists and students as well as animal breeders to accelerate animal improvement speed, to modify genetic construction of animal population, and to create new animal breeds.

# REFERENCES


1. Jarman, MR, Clark G, Grigson C, Uerpmann HP, Ryder ML, 1976: Early animal husbandry. *The Royal Soc*. 275 (936): 85-97.

2. Wolfer DP, Crusio WE, Lipp HP, 2002: Knockout mice: simple solutions to the problems of genetic background and flanking genes. *Trends Neurosci* 25(7):336-40.

3. Lai L, Kang JX, Li B, Wang J, Witt WT, Yong HY, Hao Y, Wax DM, Murphy CN, Rieke A, Sanuel M, Linville ML, Korte SW, Evans RW, Starzl TE, Prather RS, Dai Y, 2006: Generation of cloned transgenic pigs rich in omega-3 fatty acids. *Natur Biotechnol* 24: 435-436.

4. Wheeler MB, Rutledge JJ, Fischer-Brown A, Van Etten T, Malusky S, Beebe DJ, 2006: Application of sex semen technology to in vitro embryo production in cattle. *Theriogenology* 65:219-227.

5. Lillico SG, Proudfoot C, Carlson DF, Slverakova D, Neil C, Nlain C, King TJ, Richie WA, Tan, W, Mileham A, Mclaren DG, Fahrenkrug SC, Whitelaw BA, 2013: Live pigs produced from genome edited zygotes. *Sci Rep* 3:2847.

6. Seidel GE Jr, 2007: Overview of sexing sperm. *Theriogenology* 68:443-446.

7. Wu B, Zan L, 2012: Enhance beef cattle improvement by embryo biotechnologies. *Reprod Dom Anim* 47:865-871.

8. Wu B, Gelety TJ, Shi JZ, 2012: Advances in fertility options of azoospermic men. In: Bin Wu (eds.), *Advances in Embryo Transfer*. Croatia: InTech Press 2012: p. 115-132.

9. Wilmut I, Hooper ML, Simons JP, 1991: Genetic manipulation of mammals and its application in reproductive biology. *J Reprod Fert 92* 245-279.

10. Niemann H, Kues WA, 2003: Application of transgenesis in livestock for agriculture and biomedicine. *Anim Reprod Sci*. 79:291-317.

11. Greger M, 2010: Transgenesis in animal agriculture: Addressing animal health and welfare concerns. *J Agric Environ Ethics*.

12. Wheeler MB, 2003: Production of transgenic livestock: Promise fulfilled. *J Anim Sci*. 81(Suppl. 3):32–37.

13. Wheeler MB, 2007: Agricultural applications for transgenic livestock. *Trends Biotechnol* 25(5):204-210.

14. Anderson SJ, Noyes HA, Agaba M, Kemp SJ and Archibald AL, 2007: A transgenic approach to QTL analysis in a trypanotolerant mouse model. In: International Symposium on Animal Genomics for Animal Health, 23-25 October 2007, OIE HQ, World Organisation for Animal Health, 12 Rue de Prony, Paris France (unpublished).

15. Cao Z, Ding WD, 2013: Homologous rearranged DNA can change phenotype and genotype of the host by transgenic method and a QTL related to weight was obtained from it. *Adv J Food Sci Technol*. 5(3): 295-302

16. Beuzen ND, Stear MJ, Chang KC, 2000: Molecular markers and their use in animal breeding. *Vet J*. 160(1):42-52.

17. Taneja M, Yang X, 1998: Promises and problems of in vitro production of embryos by TVOR-IVF scheme in cows and heifers. *Embryo Transf Newslett.16* 10-12.

18. Machado SA, Reichenbach HD, Weppert M, Wolf E, Gonçalves PB, 2006: The variability of ovum pick-up response and in vitro embryo production from monozygotic twin cows. *Theriogenology*. 65(3): 573-583.

19. Presicce GA, Xu J, Gong GC, Moreno JF, Chaubal S, Xue F, Bella A, Senatore EM, Yang XZ, Tian XC, Du FL, 2011: Oocyte source and hormonal stimulation for *In vitro* fertilization using sexed spermatozoa in cattle. *Vet Med Int*. Published online 2010 September 5. Vet Med Int. 2011; 2011: 145626.

20. Martinez HR, 2012: Assisted reproductive techniques for cattle breeding in developing countries: A critical appraisal of their value and limitations.*Reprod Dom Anim* (47), SI, 21-26.

21. Wu B, Ignotz G, Currie WB, Yang X, 1997: Dynamics of maturation-promoting factor and its constituent proteins during in vitro maturation of bovine oocytes. *Bio Reprod* 56:253-259.

22. Thibier M, 2005: The zootechnical applications of biotechnology in animal reproduction: current methods and perspectives. *Reprod Nutr Dev*. 45:235-42.

23. Betteridge KJ, 2006: Farm animal embryo technologies: Achievements and perspectives. *Theriogenology*. 65:905-913.

24. Blondin P, Beaulieu M, Fournier V, Morin N, Crawford L, Madan P, King WA, 2009: Analysis of bovine sexed sperm for IVF from sorting to the embryo. *Theriogenology*. 71: 30-38.

25. Underwood SL, Bathgate R, Ebsworth M, Maxwell WMC, Evans G, 2010: Pregnancy loss in heifers after artificial insemination with frozen-thawed, sex-sorted, re-frozen-thawed dairy bull sperm. *Anim Reprod Sci* 118(1):7-12

26. Pontes JHF, Silva KCF, Basso AC, Rigo AG, Ferreira CR, Santos GMG, Sanches BV, Porcionato JPF, Vieira PHS, Faifer FS, Sterza FAM, Schenk JL, Seneda MM, 2010: Large-scale *in vitro* embryo production and pregnancy rates from *Bos taurus*, *Bos indicus*,

and *indicus-taurus* dairy cows using sexed sperm. *Theriogenology*. 74:1349-1355.

27. Rahbek C, 1993: Captive breeding-a useful tool in the preservation of biodiversity. *Biodivers Conserv*. 2, 426-437.

28. Hara H, Abdalla H, Morita H, Kuwayama M, Hirabayashi M, Hochi S, 2011: Procedure for bovine ICSI, not sperm freeze-drying, impairs the function of the microtubule-organizing center. *J Reprod Dev*. 57(3):428-432.

29. Abu NMAR, 2010: Intracytoplasmic sperm injection-revolution in human and animal assisted reproduction: A review. Biotechnology 9(2): 392-410.

# CHAPTER 2

# An Overview of Disease-Free Buffalo Breeding Projects with Reference to the Different Systems Used in South Africa

**Liesel Laubscher and Louwrens Hoffman** *

Department of Animal Science, University of Stellenbosch, Private Bag X1, Matieland 7602, South Africa

## ABSTRACT

This paper describes the successful national program initiated by the South African government to produce disease-free African buffalo so as to ensure the sustainability of this species due to threats from diseases. Buffalo are known carriers of foot-and-mouth disease, bovine tuberculosis, Corridor disease and brucellosis. A long-term program involving multiphase testing and a breeding scheme for buffalo is described where, after 10 years, a sustainable number of buffalo herds are now available that are free of these four diseases. A large portion of the success was attributable to the use of dairy cows as foster parents with the five-stage quarantine process proving highly effective in maintaining the "disease-free" status of both the calves and the foster cows. The projects proved the successfulness of breeding with African buffalo in a commercial system that was unique to African buffalo and maintained the "wildness" of the animals so that they could effectively be released back into the wild with minimal, if any, behavioral problems.

## KEYWORDS

foot-and-mouth disease; Corridor disease; bovine tuberculosis; brucellosis; buffalo; captive breeding

## 1. INTRODUCTION

The African buffalo is the largest of the Bovidae family and can be divided into two subspecies, namely the savanna buffalo, *Syncerus caffer caffer*, and the forest buffalo, *Syncerus caffer nananus*. The former subspecies occurs throughout Southern Africa while the latter occurs higher up throughout central and northern Africa [1]. After the 1930s, concerted efforts were made to eliminate buffalo from outside designated wildlife areas in an attempt to protect the South African cattle industry from the spread of bovine tuberculosis (BTB). As a result, most buffalo after this time appeared only in large numbers within the national parks in South Africa [2]. During the 1980s and 1990s, the savanna buffalo in South Africa became significantly threatened by the outbreak of bovine tuberculosis (BTB) and Corridor disease (CD). As a result, during the latter part of the 1990s, the South African National Parks Board initiated the first projects aimed at breeding "disease-free" buffalo. This was in an attempt to save the large gene pool of the Kruger Park buffalo, which was the most affected by these diseases. It should be noted that this was one of the main driving forces behind the initiation of "disease-free" breeding projects since a large proportion of the "disease-free" buffalo in South

Africa are of Addo Elephant Park origin. In addition, efforts were also made to contain and prevent the outbreak of brucellosis and foot-and-mouth disease (FMD) [3]. The South African government declared a veterinary redline area within South Africa in 1996 after the International Committee of the World Organization of Animal Health (OIE) declared areas in South Africa outside this zone as FMD free without vaccination [4,5]. The veterinary redline area was subsequently established in order to control the movement of buffalo and prevent the spread of FMD between infected buffalo and commercial cattle farms due to the large economic implications this held. This declaration played a major role in containing the diseases and the success of the "disease-free" breeding projects. The projects were groundbreaking, and although some break-outs of buffalo from the redline region infected with the diseases did occur before the end of such private projects in 2011, the quarantine and breeding systems implemented proved highly successful in increasing the number of buffalo in Southern Africa unaffected by FMD, BTB, CD and brucellosis.

# 2. DISCUSSION

## 2.1. A Summary of the History of the Four Most Important Diseases in Buffalo

The Cape Buffalo (*Syncerus caffer*) is a large bovid that historically roamed in large numbers across the major parts of southern Africa. They are ecologically important for their role as bulk grazers but their numbers have significantly declined in the past century due to overexploitation and disease pandemics [6]. Apart from its importance in a multispecies ecosystem, buffalo are also important from an ecotourism and economic perspective [7]. Unfortunately, 80% of South Africa's total buffalo population are carriers of one of the four major diseases that can be passed on from wild to domesticated species [8]. Although buffalo are susceptible to a number of different diseases (the most important of these will be discussed further on), foot-and-mouth disease (FMD) was the first to come under scrutiny [7]. This disease is known as an "indigenous disease" as the Southern

African Territories serotypes (SAT-1, SAT-2 and SAT-3) of the FMD virus were identified for the first time in southern Africa in the 1940s [9]. Such indigenous diseases generally do not pose a threat to wildlife populations due to the evolutionary development of unique coping mechanisms [7]. However, FMD has been under continuous scrutiny due to the economic implications of the disease, as it is passed on directly and indirectly from buffalo to cattle. The transmission of the disease is through an unknown mechanism with some authors hypothesizing its spread via aerogenous transmission under certain climatic and epidemiological conditions, while others maintain that the disease is spread through close contact between animals [10,11]. Most important, many countries do not want to import the disease via agricultural products [12]. Although the disease is not fatal, it has severe effects on South Africa's ungulate populations because of the extreme measures adopted to control it [13]. The export of agricultural products of animal origin is strictly regulated and banned if an outbreak of FMD occurs. Some authors and researchers view FMD as one of the major impediments on the continent of Africa that prevents access to lucrative export markets for animal products [12,13].

FMD is caused by an *Aphthovirus* in the family *Picornaviridae*. FMD virus (FMDV) is endemic in the Kruger National Park because a long-term carrier state exists in buffalo [13]. Research has shown that free-ranging buffalo under natural conditions are true maintenance hosts of FMDV and that infection with one of the three Southern African Territories (SAT) serotypes that cause FMDV, occurs between the early ages of three to seven months after maternally inferred immunity is diminished and almost always in the total absence of outward clinical symptoms [4,14]. After infection, the virus has been known to persist in the pharyngeal region for as long as five years with no clinical symptoms presenting as the animal then becomes a carrier of the disease [1,15,16]. Transmission between buffalo and cattle is thought to occur during close contact and usually when buffalo are in the acute phases of the disease [4,17]. Although the disease is characterized by a high morbidity (in other words, a high number of animals affected) and a low mortality (a low number of animals that actually die from the disease), an outbreak outside a controlled area may have detrimental effects on the livestock industry, especially the export of animals, animal products and agricultural products [12]. In

Europe especially, the effects of foot-and-mouth disease outbreaks have historically been great, and vast amounts of money have been spent to control the disease. As a result, the European Export Community is wary of reimporting the disease, especially exotic strains, and has instituted strict requirements on the import of animal products [2].

Since the 1970s, the testing for FMDV has been required by government for any buffalo to be moved within South Africa. At the time, other diseases were not investigated and animals that tested negative for FMDV could be moved freely between areas [3]. However, by the early 1980s, it became apparent that many of the animals infected by FMDV were also infected with Corridor disease (CD) and the disease control regulations started to change [3].

CD is one of three related diseases caused by *Theileria parva* [18]. The brown ear ticks *Rhipicephalus appendiculatus*, *R. duttoni* and *R. zambeziensis* are considered to be the main vectors of *T. parva,* which causes East Coast fever, CD and January disease in cattle [17,18]. CD results when buffalo derived *T. parva* is transmitted via ticks from buffalo to cattle, which results in mortality, although evidence has shown that cattle that survive this infection then go on to become immune to it [19,20,21,22]. Indeed, Yusufmia *et al.* (2010) [23] found evidence that suggests that cattle may also be subclinical carriers of *T. parva*. However, because in the majority of cases the disease is acutely fatal in cattle, it is considered one the most important diseases transmitted from buffalo to cattle and has contributed greatly to the need for confinement of buffalo to well-fenced areas and stringent veterinary control measures [2,17]. CD is also an indigenous disease and although the disease was first documented in Zimbabwe, it gets its name from the fact that it was recognized in South Africa in 1955 in cattle found in the corridor formed by the historic Hluhluwe and iMfolozi game parks [7,11]. Since then, recognized Corridor disease-endemic areas in South Africa include the Kruger National Park, Hluhluwe–iMfolozi Park and areas surrounding and in between these regions [19,22].

In the 1980s, *bovine tuberculosis* (BTB) was also diagnosed in buffalo in the Hluhluwe-iMfolozi Park, which presented a new threat to the game and agricultural industry [24]. In June 1990, the first buffalo bull in Kruger National Park was diagnosed with BTB, which prompted an investigation into the BTB status of all Kruger Park buffalo [25]. By the late 1990s, numerous buffalo herds in both the Kruger National Park and the Hluhluwe–iMfolozi Park were found to be infected with BTB and it is thought that the disease was introduced by imported European cattle breeds mainly during the 18th and 19th centuries [24,25]. Subsequent outbreak investigations indicated that the disease was likely to have been transmitted from domesticated cattle to buffalo in the Kruger National Park in the southeast corner of the park between 1950 to 1960 [2,23,24]. BTB is therefore known as a foreign, so-called alien livestock disease which is caused by the bacterium *Mycobacterium bovis*and can spread quite rapidly from buffalo to cattle and *vice versa*. As it had previously been introduced into the country, it had successfully been established in buffalos which then served as a wildlife maintenance host [7,23]. Once an animal is infected, it will spread spontaneously through the herd when cough droplets from infected animals are inhaled by healthy animals as well as through food and water contamination [1]. Although infected buffalo have been known to live with BTB for several years before clinical signs develop, the disease is fatal and has been seen to spill over into other wildlife species such as baboon, lion, cheetah and kudu when it reaches a certain level within a buffalo herd. It has been described as a chronic, progressively debilitating condition [25].

Another disease found in African buffalo that is transferable to livestock and that is now common practice to test for is bovine brucellosis, which is also classified as a foreign livestock disease [7]. It is caused by the bacterium *Brucella abortus*, one of six species recognized within the genus *Brucella* [26,27]. It is usually transmitted orally via contact with fetal membranes, post-parturition discharges and milk as well as through the semen of infected males [26,27,28]. The disease causes abortion in the latter half of gestation and is also known as "contagious abortion." This may result in massive economic losses, as well as pose a potential barrier to international trade [29]. The disease presents with very noticeable outward symptoms as it leads to highly swollen carpal joints, known as "hygromata" or "house maid knees syndrome." In South Africa, no breeding is allowed with buffalo that have tested positive for brucellosis and aborted fetuses have to be handled with care as the disease is transmissible to humans [1].

## 2.2. The History of Disease-Free Buffalo Breeding Projects in South Africa

In 1996, the South African National Parks Board (SANPARKS) requested permission for and started the first official pilot studies for breeding "disease-free" buffalo. Part of the reason was that SANPARKS wanted to increase "disease-free" buffalo numbers because many buffalo herds had gone extinct from their former ranges due to over-hunting, the rinderpest epidemic (caused by the Rinderpest virus or RPV) and exclusion from cattle ranching areas. They also aimed to preserve the most heterogeneous buffalo genetics from other parks other than Kruger Park [30]. These pilot studies were designed to produce "disease-free" calves from "diseased" parent stock to provide official conservation organizations and the wildlife ranching industry with clean "disease-free" buffalo from controlled areas that did not pose any disease risk [6]. The studies made use of diseased parent stock from the Kruger National Park as it had the largest gene pool within its herds and the aims were to preserve this buffalo genotype if ever a depopulation decision was made to control BTB as well as to supply other parks with "disease-free" buffalo of the Kruger genotype [3]. There were two pilot studies approved, one that left calves with their diseased mothers until their maternally derived immunity vanished and a privately funded project where calves were taken from their mothers at birth and fostered by surrogate dairy cows. After the success of these pilot studies, numerous other similar projects were launched and, in 1998, a formal Buffalo Advisory Committee was formed to control disease-free buffalo breeding projects through official registration of such projects as well as the monitoring of movement of animals from controlled to noncontrolled areas, among other things [3]. In 2001, a meeting was held by the Committee in Hoedspruit after a number of "breakthrough" infections indicated that the breeding projects did not sufficiently lower the risk of diseases spreading. At this meeting, it was decided that these disease-free buffalo breeding activities would be phased out by

December 2011 as by such time, enough “disease-free” stock should have been produced. Before the projects were phased out, there were 28 official “disease-free” buffalo breeding projects in South Africa [6]. As the demand for “disease-free” buffalo increased, so did the price of these animals and many farmers started such breeding projects for the potentially large economic benefits they held [31]. Today there are over 130 registered farms (defined as farms containing only buffalo that are “disease-free”) in the province of Mpumalanga, South Africa alone. A comprehensive and complete list of farms registered within South Africa as “disease-free” buffalo farms can be found on the South African Department of Agriculture, Forestry and Fisheries website under Disease Control under Food and Veterinary Services [8].

## 2.3. The “Disease-Free” Breeding System

In their natural environments, breeding would occur in accordance with the availability of nutritionally adequate feed [32]. However, in captivity, buffalo breed throughout the year although most births occur during the summer months [33]. Similar to any other agricultural animal breeding practice, the breeding stock for a “disease-free” buffalo breeding herd is chosen based on specific principles and standards. Bulls are selected on temperament and semen quality as well as horn length, the latter being an important economical factor for the hunting industry as many hunters select trophy animals based on this trait. Cows are selected primarily based on fertility and both cows and bulls have to be free of BTB and brucellosis to be selected. The breeding stock can either be purchased directly from farmers, at wildlife auctions or through reputable wildlife capture operators. “Diseased” stock can be as much as a third of the price of “disease-free” stock because of the large financial costs involved in using such stock to breed “disease-free” calves [1]. When “diseased” stock are captured directly from the field, the test for brucellosis is done in the field while the animals are immobilized and animals that test positive are released. Thereafter, animals are tested while in captivity for BTB (using the intradermal skin test) and animals that test positive are euthanized. If any animals test positive, the entire herd captured is rejected [30].

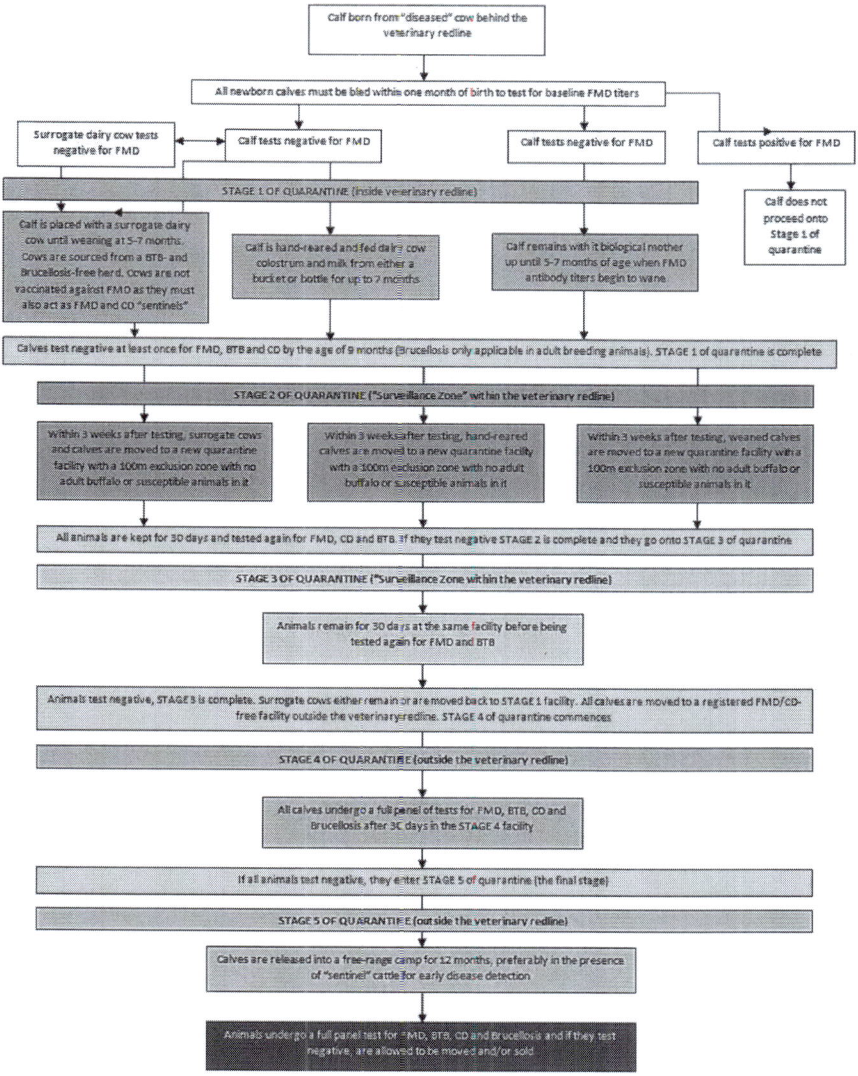


**Figure 1.** The different breeding systems used to breed "disease-free" buffalo calves from "diseased" parent stock.

There are three basic systems that are used to obtain the "disease-free" calves, each system with its own advantages and pitfalls (refer to Figure 1). The first two systems both make use of either foster dairy cows to act as surrogate mothers to buffalo calves or of feeding the calves with bottles or buckets. All three systems make use of intensive breeding practices so that breeding cows are kept in sheds built according to exact specifications, each with at least one quarantine station (*i.e.*, a box within a box). Such intensive breeding systems make it possible for the breeder to remove calves from their mothers at specific times, depending on the breeding system implemented [1,25].

In the first system, the calves are taken away immediately from the buffalo cow after birth and fed cattle colostrum. Thereafter, it is either hand-reared or placed with a surrogate dairy cow. The benefits that this system holds include a severe decrease in the risk of FMDV (which may be contracted once the colostrum-derived FMDV antibody titers begin to wane after several months) and CD being contracted by the calf [27]. Even if ticks manage to break through the tick-control efforts, they are unlikely to carry the *T. parva* parasite as there would be no animals with CD in the surrounding areas. The calves also tend to become more tame and easier to handle, and no lactational suppression of estrus results in a shorter calving interval for the buffalo cow. The latter fact is supported by Skinner *et al.* (2006), who reported more peaks in percentage calves born in captivity and attributed this to both the high nutritional value of the feed given as well as the fact that calves were taken away immediately from their mothers [32]. However, this system also holds the disadvantage of being highly labor and management intensive as the early detection of parturition is critical and calves need to be given good quality cattle colostrum at the correct dosage to ensure adequate early disease resistance [27]. Additionally, although the calves are tamer, they are still dangerous and a number of incidents with calves/young animals attacking humans have been noted.

The second system makes use of the same principles as the first, except that calves are allowed to drink the colostrum naturally produced by their mothers before being taken away. This means the calf stays with its mother for roughly the first 48 hours of its life before being removed and either hand-reared using a bucket or bottle, or is placed with a foster cow. Making use of this method means that calves will have positive titers to FMDV for a variable period, thereby delaying the completion of stage 1 quarantine and their removal date from the infected zone facility. However, it also ensures that calves gain protection against diseases from the buffalo's colostrum as well as learning to suckle strongly before being taken away [27].

The choice between the first two methods is often the personal preference of the breeder, and calves are taken away from their mothers by pushing the mothers into a different pen and closing the gate before the much slower calf can catch up. If this method is unsuccessful, the mother may also be tranquilized via a projectile dart so that the calf can be removed. Once the calf is removed, it is washed and drenched with either pyrethroids or formamidines for treatment against any external parasites [22]. The calf is also weighed, the sex determined, and it is given an ear tag and microchip for identification. All this data is kept as part of a calf register for future use. Calves that have been removed from their mothers are then placed in groups of five to eight animals and either hand-reared or fostered by dairy cows at a ratio of two to three calves per cow. Jersey cows are the preferred foster mothers, due to their good mothering abilities and high milk yield [1].

Using both the aforementioned systems, breeders also have the option to hand-rear calves that have been removed from their diseased mothers. In this case, the calves are hand-reared on whole cow's milk from registered disease-free dairies using either buckets or bottles. A disadvantage of doing this is that it is highly labor intensive and calves may imprint on the humans feeding them as they tend to become tame quite easily, resulting in aberrant behavior later in life. This behavior may only become apparent once the grown calves are released into a larger herd as they may become aggressive and dangerous due largely to no inherent fear of humans [27].

The alternative is to place the calves with foster dairy cows, which is often the preferred method used by breeders (Figure 2). The dairy cows are selected based on a number of traits. Firstly, the cow has to be in good health, and be BTB- and brucellosis-free, along with other health traits such as no history of mastitis, strong feet, *etc.* To prevent mastitis from occurring, it is generally recommended that cows be selected with a peak milk production of less than 25 liters per day. Buffalo milk has a butterfat content of 6–8% [34], therefore, it is recommended to select cows with a high butterfat production rate so as to compensate for the replacement of buffalo milk. Also, buffalo calves nurse between the mother's hind legs so cows need to be selected with long teats to assist the calf in suckling. Jersey cows tend to make good surrogate mothers and seldom reject calves that attempt to suckle from behind [35]. Cows that have had their second parturition tend to make better foster mothers, as well, as they are more comfortable and less nervous than first-time mothers because they have already been handled in the milking parlor. Lastly, since cows have to be tested every three months and are likely to be handled on a daily basis, it is important that they are also selected for temperament, so as to ease handling procedures and avoid injuries to calves and workers [1].

**Figure 2.** A buffalo calf with a foster jersey dairy cow. The calf has just been darted to draw blood for testing.

An alternative to hand-rearing and using surrogate dairy cows to raise buffalo calves is a third system whereby calves are left with their biological mothers, who originate from potentially "diseased" stock, after birth. They therefore obtain immunity naturally from buffalo colostrum as well as being exposed to normal buffalo herd behavior so that behavioral problems later on in life are prevented [27]. However, this system is seldom recommended as it holds numerous disadvantages. Studies suggest that the FMDV virus is present in the semen of bulls so that there is the possibility that it may be passed on to the cow during mating and then onto the calf through the milk if the mother was mated while still lactating [1,36]. Calves also tend to grow slower than their counterparts raised on Jersey milk and, as mentioned before, the intercalving period for the cows become longer if their calves are kept with them because of lactation anestrus. Also, as the calf's passive immunity wanes, there is a finite risk of a breakthrough FMDV infection, as well as the risk of a *Theileria*infection if there is a breakdown in the tick control, since calves will still be in the presence of their *Theileria* infected mothers [27]. Lastly, if conditions are overcrowded and often wet, outbreaks of diseases such as *Escherichia coli* and coccidiosis that lead to diarrhea and mortality in young calves may occur [1]. Calves that are left with their biological mothers are usually weaned betweenthe ages of five and seven months, or at 100 kg of weight when the colostrums-derived FMDV antibody titers have started to wane significantly. Prior to this, this maternally derived immunity prevents calves from becoming infected with FMDV [30]. However, fostered calves may be weaned earlier before they start to grow horns and become big and strong enough to injure their jersey surrogates [27].

### 2.3.1. Quarantine Procedures for Breeding of "Disease-Free" Buffalo

The veterinary redline area (infected zone) in South Africa is classified as the area in and around the Kruger National Park, including Mozambique and a 10- to 15-km inspection zone bordering Swaziland, Zimbabwe and Botswana (free movement of animals from the inspection zone to the free zone is prohibited) where FMDV is most prevalent (Figure 3) [1,34]. This control area is defined by natural geographical borders, roads and fences [37,38]. Adjacent to the western and southern borders of this infected zone is the buffer zone, which has two sections: a portion where animals are vaccinated twice yearly (referred to as the buffer zone with vaccination (BZV)) and a portion where animals are not vaccinated but where increased surveillance and movement control are implemented, known as the buffer zone without vaccination (BZNV)) [4]. Buffalo that occur behind the redline and that are infected with one of the four diseases previously discussed, are often referred to as "dirty buffalo" and movement of such buffalo is strictly regulated to prevent diseases spreading to unaffected animals [39]. However, diseased animals inside the redline area that test negative for brucellosis and BTB are used as breeding stock in "disease-free" breeding projects. While using these latter animals for breeding, they are all kept in quarantine and the property must fall behind the veterinary redline as it is illegal to move animals from behind the redline to outside the redline area unless they have been bred as "disease-free" following the strict protocol discussed below [39].

All parent stock must test negative for BTB and, if they are sourced from a unknown BTB area, all animals must undergo an additional two tests for BTB after the initial three-month quarantine period before they will be considered BTB-free and allowed to join a BTB-free herd for breeding. If animals are sourced from a known BTB area, they have to undergo at least five negative BTB tests over a period of fifteen months before being considered BTB-free and the first of these tests can only be done three months post-capture [39]. As mentioned, all breeding animals must also test negative for brucellosis, and cows have to test negative at least three times, as well as be tested once a year thereafter, usually two to twelve weeks postcalving. Foster dairy cows may not be vaccinated against FMDV so that they can fulfill their additional roles as "sentinels" for the disease [27].

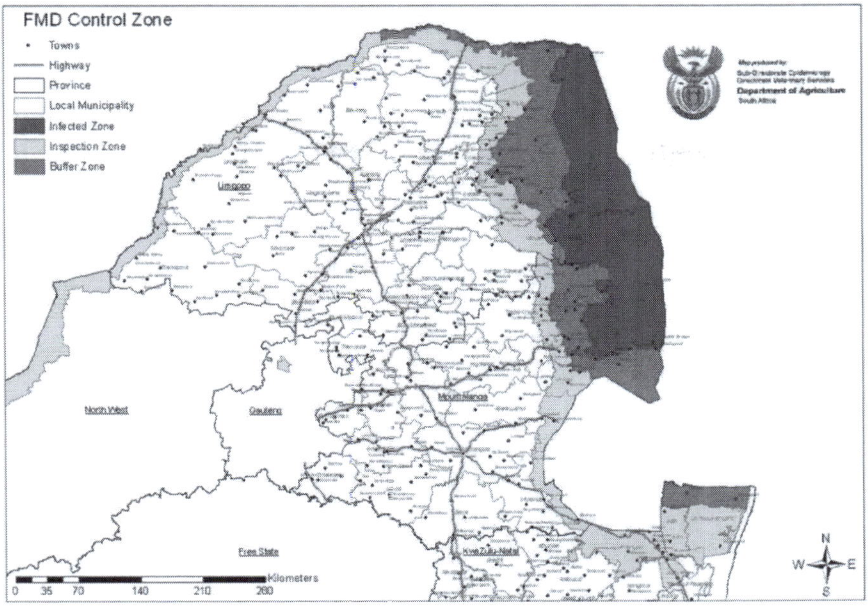


**Figure 3.** Disease control zones in South Africa according to the South African Department of Agriculture, Forestry and Fisheries [8].

When a calf is born from a FMDV or a potential FMDV mother, it is taken away within 24 hours for testing, sexing and tagging (Figure 1). This first testing must be performed at least 100 m from the "diseased" parents and all surrogate mothers must be tested parallel to the calves. If the calf is taken away permanently from its mother, it has to be bled within one month for baseline FMDV titers and the results will indicate whether the calf ingested any of its mother's colostrum. If the test indicated no buffalo colostrum was ingested, then stage 1 of the quarantine process begins and calves also have to test negative at least once for Corridor disease and BTB. Once they do, stage 1 of the quarantine is complete [30]. Those calves that have ingested or been allowed to drink their biological mother's colostrum, must be bled at five to seven months of age (when they are weaned) to test for residual FMDV titers. If they test positive, this is not considered an outbreak as the occurrence falls within the veterinary redline area. Financial implications due to the low marketability of such calves are carried by the commercial operator. However, if they test negative then they also have to be tested for Corridor disease and BTB, and once they test negative for all three diseases, stage 1 of their quarantine is complete. On completion of stage 1, all calves and surrogate cows are moved within three weeks of sampling to a quarantine facility within the FMDV surveillance zone under the supervision of the local state veterinarian. All the calves and foster cows now have the same disease status and the quarantine facility will have no adult buffalo in it and will be surrounded by a 100 m exclusion zone with no susceptible animals in it. This is the second stage of quarantine and all animals will remain at this facility for 30 days and all calves and foster cows will be bled again for FMDV, BTB and Corridor disease testing. If they test negative, they enter stage 3 of quarantine and must remain for a further 30 days in the same facility before being bled again for FMDV and BTB. Once they test negative again, stage 3 is completed and foster cows either remain in the surveillance zone facility or are moved back to the primary facility in the infected zone while the calves are moved under the supervision of the state veterinarian to a suitable facility that is registered as FMDV/Corridor disease free in an area outside the veterinary redline [39]. This is the fourth stage of quarantine for the buffalo calves and after 30 days at the new facility, all the calves have to be re-bled for a full panel test for FMDV, BTB, Corridor disease and brucellosis. This last testing must occur before the calves are 11 months old and once they test negative, they enter the fifth and final stage of quarantine. At this stage, calves are released into a free-range camp for a retention period of 12 months during which time they are exposed to brown-ear ticks and are preferably in the presence of FMDV "sentinel" cattle, which act as an early detection system for the disease [1,35]. The sentinel cattle are much tamer than the buffalo and thus require no specialized equipment for immobilization when blood sampling is done and are thus far easier to test for disease status in the absence of outward symptoms. After this 12-month period, all animals undergo a full panel test again for all four diseases and, if they test negative, are allowed to be moved/sold under the supervision of the state veterinarian. In Corridor disease-only breeding projects in the vector-free areas, the last two stages of quarantine do not apply if animals are destined to be sold to farms within the vector-free area [1,25,28].

Disease-free buffalo are thereafter traded freely on the South African game species market and make up a large percentage of the profit margin at annual game auctions [40].

### 2.3.2. Testing Procedures

Diagnostic testing of infectious diseases may often be difficult in wildlife if they are taxonomically far removed from livestock. However, buffalo fall within the subfamily Bovidae, which thus provides a basis for the adaptation and development of diagnostic tests for BTB, FMDV, CD and Brucellosis. Although a number of diagnostic tests are currently being used to test for these diseases, these tests still require validation to prove fitness of purpose as stated by the World Organization for Animal Health (OIE) [7].

#### *2.3.2.1. Bovine Tuberculosis*

For BTB, the standard method of detection is to inject a small amount of the antigen into the skin of the animal and then to measure the immune response. This is usually done by measuring skin thickness 72 hours postinjection to determine if any swelling has occurred (Figure 4). This method is also known as the comparative intradermal test, or skin test [27,41]. This test is used as the golden standard in South Africa but may lose some accuracy if there is a high level of environmental exposure to other sensitizing organisms. According to the World Organization for Animal Health (OIE), purified protein derivatives (PPD) products have replaced heat-concentrated synthetic medium tuberculins because of their higher specificity and easier standardization, and no less than 2000 International Units (IU) of bovine PPD should be used when performing the skin test [42].

#### *2.3.2.2. Brucellosis*

In the case of brucellosis, any clinical signs such as abortion or vaginal discharge should be treated as suspicious. A number of serological tests are suitable for detection of *Brucella*, including the Rose Bengal test and buffered plate agglutination test, the complement fixation test (CFT), the enzyme-linked immunosorbent assay (ELISA) or the fluorescence polarization assay [43]. However, no single test is appropriate in all epidemiological situations and, in African buffalo, the complement fixation test is considered suitable, although Rose Bengal reactions are also treated with suspicion [27].

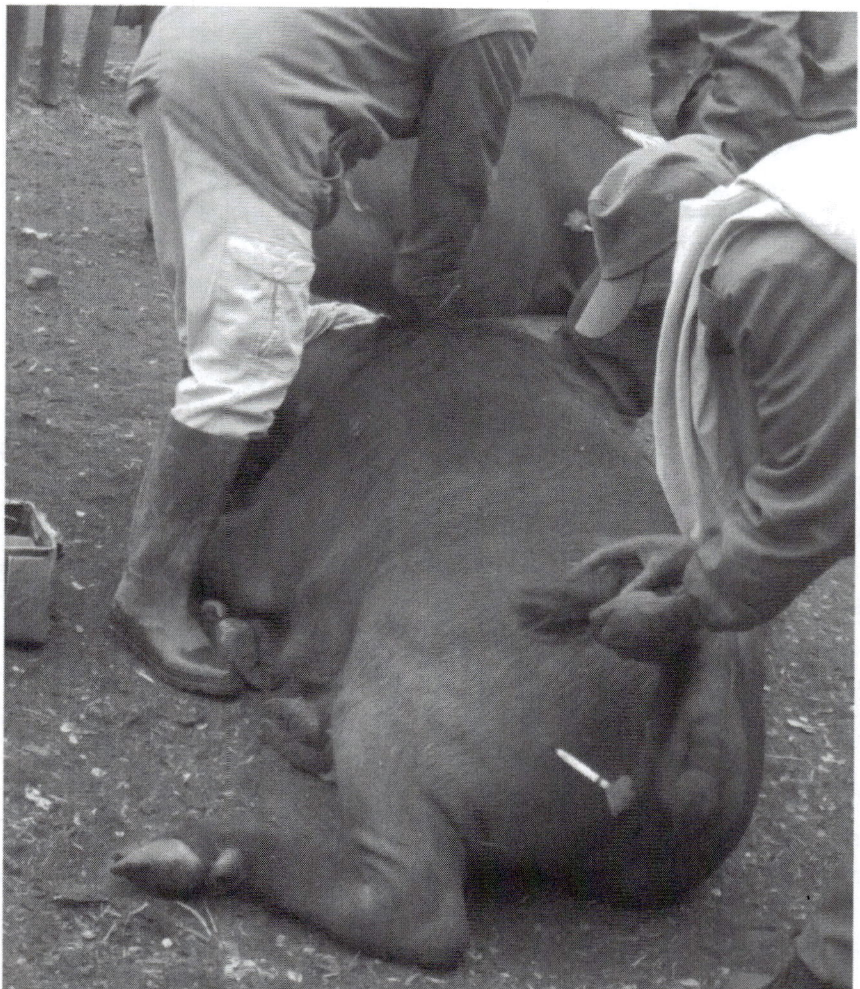

**Figure 4.** Immobilized buffalo being tested for BTB using the skin test.

*2.3.2.3. Foot-and-Mouth Disease*

When an animal is infected with FMDV, its body produces both structural proteins (SP) and nonstructural proteins (NSP) of the virus. SP antibodies can be tested for with SP-ELISA's or the virus neutralization test (VNT), the latter of which depends on tissue cultures and are therefore more prone to variability than ELISAs. VNT's are also slower and subject to contamination [44]. NSP antibodies can be tested for with NSP I-ELISA 3ABC and the electro-immunotransfer blotting technique (EIBT). Whereas the SP tests are serotype specific, the NSP tests can detect antibodies to all serotypes of FMDV [45].

*2.3.2.4. Corridor Disease*

To test if an animal is infected with Corridor disease, the animal must be tested using the indirect immunofluorescent antibody (IFA) test or polymerase chain reaction (PCR)/probe assay. Disadvantages of the IFA test include cross-reactivity between different species, subjectivity in the interpretation of the results and difficulty in standardization [46]. If animals are not subject to a continuous tick challenge, it may also happen that the antibodies are not detected and as a result a number of other techniques have also been developed for detection of CD using PCR and DNA probes. Smear examination may also be used, although this can only detect *Theileria* piroplasms and can as such not be used to make a completely accurate diagnosis of CD [27,46].

### 2.3.3. Rationale Behind the Extensive Protocol Used for "Disease-Free" Breeding Projects

The purpose of this extensive protocol is to ensure that the buffalo are free of the four diseases: FMD, CD, TB, and brucellosis. The test for FMD is very specific and very sensitive. However, it does not distinguish between maternal immunity and acquired immunity. Research has proven that the maternal immunity will have completely dissipated after nine months, hence the animals have to stay in quarantine for this period for the sake of FMD. The brucellosis is also sensitive and specific and because no positive maternal animals are allowed to participate, it is almost guaranteed that this test is always negative. TB, however, has a less sensitive and specific test and, therefore, the test is only considered as a herd test and not an individual test. For this reason, the protocol requires five negative tests of which the first four is taken at three-month intervals, and the last test only a year later. This will hopefully pick up any new infections should there be TB in the animals. For CD, the test is extremely sensitive, but not that specific. CD is tick transmitted and, before the last test, the animals must be exposed to the vectors for a period of one year, which will allow time for infection of all animals should the disease be present. There has been no definitive proof of any transplacental infections for any of these diseases. Unfortunately, calves are removed from the program the minute they test positive for any of these diseases, and their outcomes are unknown [47].

## 3. CONCLUSIONS

Since African buffalo came under the threat of FMDV, BTB, Corridor disease and brucellosis, the breeding of disease-free buffalo became an increasingly lucrative business [48]. The average price for a "disease-free" buffalo in 2004 was almost ZAR150,000 (ZAR1 ≈ US$0.12) with prices increasing by an average of 29% per year as the demand for "disease-free" buffalo increased [29,34]. In 2008, "disease-free" buffalo were being sold for just over ZAR160,000, while in 2010, this price increased to over ZAR325,000 per buffalo at the same auctions [49]. Towards the end of 2011, an amount of ZAR18 million was paid for a buffalo bull, while in 2012, ZAR20 million was paid for a buffalo cow and her bull calf, as well as a record price of R26 million for a buffalo bull [50]. As a result, a large number of "disease-free" breeding projects were initiated solely for economic reasons although the primary reason for initiating such projects was to maintain and propagate the Kruger Park buffalo gene pool, which had been severely affected by FMDV, BTB, brucellosis and Corridor disease and to increase the distribution of "disease-free" buffalo across South Africa. In 2011, the South African government put an end to the state-funded "disease-free" buffalo breeding projects outside of the national parks [6]. Part of the reason for ending the national "disease-free" breeding projects was the fact that sufficient privately owned breeding herds that could function as nuclei for repopulation of the large herds in the national parks, if ever required, had been established, and the government had no wish to compete economically with the private sector. While the projects were running, they were highly effective in increasing the number of "disease-free" buffalo in South Africa and maintaining and spreading the large gene pool of the Kruger National Park buffalo. However, a number of outbreaks were still reported between the start of the projects using diseased breeding stock in 2011, which further spurred the government's decision to end state-funded breeding projects. However, these outbreaks should not divert from the success of the projects as a large number of these were directly linked to escapees from the national parks rather than from the breeding projects themselves [3,19].

The pilot studies for these projects were groundbreaking and led to the successful use of diseased buffalo stock as breeding parents for "disease-free" buffalo calves. A large portion of the success was attributable to the use of dairy cows as foster parents with the five-stage quarantine process, proving highly effective in maintaining the "disease-free" status of both the calves and the foster cows. The projects proved the successfulness of breeding with African buffalo in a commercial system that was unique to African buffalo and maintained the "wildness" of the animals so that they could effectively be released back into the wild with minimal, if any, behavioral problems.

## CONFLICT OF INTEREST

The authors declare no conflict of interest.

# REFERENCES


1 Du Toit, J.G. Disease free buffalo. Available online: http://wernerbooysen.com/werdocs/BUFFALO_01.pdf (accessed on 11 July 2012).

2 Winterbach, H.E.K. Status and distribution of Cape buffalo (*Syncerus caffer*) in South Africa. *S. Afri. J. Wildl. Res.* **1998**,*28*, 82–88.

3 Saayman, R. Buffalo farming and breeding projects. Available online: http://gameranger.co.za/wp-content/uploads/2011/01/Buffalo-Breeding-Projects-Game-Ranchers-Association-Part-1.pdf (accessed on 11 July 2012).

4 Jori, F.; Vosloo, W.; Du Plessis, B.; Bengis, R.; Brahmbhatt, D. A qualitative risk assessment of factors contributing to foot and mouth disease outbreaks in cattle along the western boundary of the Kruger National Park. *Rev Sci Tech.***2009**, *28*, 917–931.

5 South African National Department of Agriculture. Submission to the foot and mouth disease and other epizootics commission of the OIE for the re-instatement of a foot and mouth disease free zone without vaccination. Available online: http://www. nda.agric. za/docs/GenPub/fmdreport.htm#1.2 (accessed on 5 June 2012).

6 South African Department of Agriculture Forestry and Fisheries. Publication of proposed notice on measures to control certain animal diseases. *Staatskoerant No. 32944*; South African Department of Agriculture Forestry and Fisheries: Pretoria, South Africa, 2010; pp. 3–4.

7 Michel, A.L.; Bengis, R.G. The African buffalo: A villain for inter-species spread of infectious diseases in southern Africa. *Onderstepoort J. Vet.* **2012**, *79*, 1–5.

8 South African Department of Agriculture Forestry and Fisheries Food Safety and Veterinary Service. Available online:www. daff. gov.za (accessed on 5 June 2012).

9 Brown, F. The history of research in foot-and-mouth disease. *Virus Res.* **2003**, *91*, 3–7.

10 Donaldson, A.I.; Alexandersen, S. Predicting the spread of foot and mouth disease by airborne virus. *Rev. Sci. Tech.***2002**, *21*, 569–575.

11 Vosloo, W.; Bastos, A.D.; Sangare, O.; Hargreaves, S.K.; Thomson, G.R. Review of the status and control of foot and mouth disease in sub-Saharan Africa. *Rev. Sci. Tech.* **2002**, *21*, 437–449. [Google Scholar]

12 Dawe, P.; Sorensen, K.; Ferris, N.; Barnett, I.; Armstrong, R.; Knowles, N. Experimental transmission of foot-and-mouth disease virus from carrier African buffalo (*Syncerus caffer*) to cattle in Zimbabwe. *Vet. Rec.* **1994**, *134*, 211–215.

13 Meltzer, D.G. Historical survey of disease problems in wildlife populations: Southern Africa mammals. *J. Zoo Wildl. Med.* **1993**, *24*, 237–244.

14 Vosloo, W.; de Klerk, L.; Boshoff, C.I.; Botha, B.; Dwarka, R.M.; Keet, D.; Haydon, D.T. Characterisation of a SAT-1 outbreak of foot-and-mouth disease in captive African buffalo (I): Clinical symptoms, genetic characterisation and phylogenetic comparison of outbreak isolates. *Vet. Microb.* **2007**, *120*, 226–240.

15 Moerane, R.; Penrith, M.L., Sibanda, R.; Thomson, G. Foot-and-mouth disease and market access: Challenges for the beef industry in southern Africa. *Pastoralism* **2010**, *1*, 135–164.

16 Candy, J.B.; Hedger, R.S. Experiences in the establishment of a herd of foot-and-mouth disease free African buffalo. *S. Afr. J. Wildl. Res.* **1978**, *8*, 87–89.

17 Condy, J.; Hedger, R.; Hamblin, C.; Barnett, I.T. The duration of the foot-and-mouth disease virus carrier state in African buffalo (i) in the individual animal and (ii) in a free-living herd. *Comp. Immunol. Microb.* **1985**, *8*, 259–265.

18 Norval, R.A.; Lawrence, J.A.; Young, A.S.; Perry, B.D.; Dolan, T.T.; Scott, J. Theileria parva: Influence of vector, parasite and

host relationships on the epidemiology of theileriosis in southern Africa. *Parasitology* **1991**, *102*, 347–356.

19 Potgieter, F.T.; Stoltsz, W.H.; Blouin, E.F.; Roos, J.A. Corridor disease in South Africa: A review of the current status. *J. S. Afr. Vet. Assoc.* **1988**, *59*, 155–160.

20 Perry, B.D.; Kruska, R.; Lessard, P.; Norval, R.A.I.; Kundert, K. Estimating the distribution and abundance of*Rhipicephalus appendiculatus* in Africa. *Prev. Vet. Med.* **1991**, *11*, 261–268.

21 Pienaar, R.; Potgieter, F.T.; Latif, A.; Thekisoe, O.M.M.; Mans, B.J. Mixed *Theileria* infections in free-ranging buffalo herds: Implications for diagnosing *Theileria parva* infections in Cape buffalo (*Syncerus caffer*). *Parasitology* **2011**, *24*, 1–12.

22 Torr, S.; Eisler, M.; Coleman, P.; Morton, J.; Machila, N. *Integrated Control of Ticks and Tsetse; A Report for the DFID Advisory and Support Services Contract*; DFDID: London, UK, 2001; pp. 1–150.

23 Yusufmia, S.B.A.S.; Collins, N.E.; Nkuna, R.; Troskie, M.; Van den Bossche, P.; Penzhorna, B.L. Occurrence of *Theileria parva* and other haemoprotozoa in cattle at the edge of Hluhluwe-iMfolozi Park, KwaZulu-Natal, South Africa. *J. S. Afr. Vet. Assoc.* **2010**, *81*, 45–49.

24 Neitz, W.O.; Canham, A.S.; Kluge, E. Corridor disease: A fatal form of bovine theileriosis encountered in Zululand. *J. S. Afr. Vet. Assoc.* **1955**, *26*, 79–87.

25 De Vos, V.; Bengis, R.G.; Kriek, N.P.; Michel, L.; Keet, D.F.; Raath, J.P.; Huchzermeyer, H.F.J.K. The epidemiology of tuberculosis in free-ranging African buffalo (*Syncerus caffer*) in the Kruger Park, South Africa. *Onderstepoort J. Vet.* **2001**, *68*, 119–130.

26 Michel, L.; Bengis, R.G.; Keet, D.F.; Hofmeyr, M.; De Klerk, L.M.; Cross, P.C.; Jolles, E.; Cooper, D.; Whyte, I.J.; Buss, P.; Godfroid, J. Wildlife tuberculosis in South African conservation

areas: Implications and challenges. *Vet. Microb.***2006**, *112*, 91–100.

27 South African National Department of Agriculture. *Disease Risk Management for Buffalo* (Syncerus caffer) *in South Africa*; Draft Number 20 (2002/09/19). National Department of Agriculture: Pretoria, South Africa, 2002; pp. 1–27.

28 Chaparro, F.; Lawrence, J.V.; Bengis, R.; Myburgh, J.G. A serological survey for brucellosis in buffalo (*Syncerus caffer*) in the Kruger National Park. *J. S. Afr. Vet. Assoc.* **1990**, *61*, 110–111.

29 Godfroid, J. Brucellosis in wildlife. *Rev. Sci. Tech.* **2002**, *21*, 277–286.

30 Hofmeyr, M. Disease-Free Buffalo Project of the State Veterinary Services and South African National Parks. In IUCN 2003 World Parks Congress AHEAD Launch Forum, Durban, South Africa, 14–15 September 2003; pp. 1–11.

31 Van der Merwe, N.; Saayman, M. Factors that determine the price of game. *Koedoe* **2010**, *47*, 105–113.

32 Skinner, J.D.; Dott, H.M.; Matthee, A.; Hunt, L. Captive breeding of the white rhinoceros, Ceratotherium simum, and the Cape buffalo, *Syncerus caffer*. *Onderstepoort J. Vet.* **2006**, *73*, 237–239.

33 Skinner, J.D.; Moss, D.G.; Skinner, D.C. Inherent seasonality in the breeding seasons of African mammals: Evidence from captive breeding. *T. Roy. Soc. S. Afr.* **2002**, *57*, 25–27.

34 Osthoff, G.; Hugo, A.; De Wit, M.; Nguyen, T.P.M. The chemical composition of milk from free-ranging African buffalo (*Syncerus caffer*). *S. Afr. J. Wildl. Res.* **2009**, *39*, 97–102.

35 Estes, R.D. *The Behavior Guide to African Mammals*; Russel Friedman Books CC, Halfway House: Johannesburg, South Africa, 1997.

36 Bastos, A.D.S.; Bertschinger, H.J.; Cordel, C.; Vuuren, C.D.W.J.V.; Keet, D.; Bengis, R.G.; Grobler, D.G.; Thomson,

G.R. Possibility of sexual transmission of foot-and-mouth disease from African buffalo to cattle. *Vet. Rec.* **1999**, *145*, 77–79.

37 Hay, C.T. Choice of social environment of male buffalo (*Syncerus caffer*) in the Kruger National Park. Mtech Thesis, Tswane University of Technology, South Africa, 2006; pp. 1–106.

38 Brückner, G.K.; Vosloo, W.; Du Plessis, B.J.; Kloeck, P.E.L.G.; Connoway, L.; Ekron, M.D.; Weaver, D.B.; Dickason, C.J.; Schreuder, F.J.; Marais, T.; Mogajane, M.E. Foot and mouth disease: The experience of South Africa. *Rev. Sci. Tech.* **2002**, *21*, 751–764.

39 South African National Department of Agriculture and National Directorate of Animal Health. *Foot-and-Mouth Disease Control Movement Protocol*; Department of Agriculture, Forestry and Fisheries: Pretoria, South Africa, August 2005.

40 Damm, G. Game sale statistics South Africa 2002 to 2004. *African Indabe e-Newsletter* **2005**, *3*, 16–18.

41 Bernardelli, A.; Cousins, D.V. Bovine Tuberculosis. Available online: http://www.oie.int/fileadmin/Home/eng/Media_Center/doc s/pdf/Disease_cards/BOVINE-TB-EN.pdf (accessed on 11 July 2012).

42 World Organization for Animal Health. Bovine Tuberculosis. Available online: http://www.oie.int/ fileadmin/Home/ eng/ Health_standards/tahm/2.04.07_BOVINE_TB.pdf (accessed on 4 October 2012).

43 World Organization for Animal Health. Bovine Brucellosis. Available online: http://www.oie.int/fileadmin/ Home/eng/Health_ standards/tahm/2.04.03_BOVINE_BRUCELL.pdf (accessed on 4 October 2012).

44 World Organization for Animal Health. Foot and mouth disease. Available online: http://www.oie.int/eng/A_ FMD2012/docs/ 2.01.05_FMD.pdf (accessed on 4 October 2012).

45 World Organization for Animal Health. Foot and mouth disease. Available online: http://www.oie.int/en/international-standard-

setting/terrestrial-code/access-online/?htmfile= sommaire.htm (accessed on 4 October 2012).

46 Sibeko, K.P. Improved molecular diagnostics and characterization of *Theileria parva* isolates from cattle and buffalo in South Africa. PhD Thesis, University of Pretoria, South Africa, 2009; pp. 39–48.

47 Raath, J.P. Wildlife Pharmaceuticals SA. Personal communication, 2012.

48 Weaver, L.C.; Skyer, P. Conservancies: Integrating wildlife land-use options into the livelihood development and conservation strategies of Namibian communities. In 5th World Parks Congress, Durban, South Africa, 2003; pp. 1–40.

49 Cloete, F. Tendense van lewende wildverkope. *Wild & Jag* **2011**, 41–43.

50 Christie, S. Wildlife prices too big for genes. Available online: http://mg.co.za/article/2012-04-20-wildlife-prices-too-big-for-genes (accessed on 7 June 2012).

# CHAPTER 3

# The lore of low methane livestock: co-producing technology and animals for reduced climate change impact

**Ann Bruce**

ESRC Innogen Centre, Old Surgeons' Hall, High School Yards, The University of Edinburgh, Edinburgh EH1 1LZ, UK

## ABSTRACT

Methane emissions from sheep and cattle production have gained increasing profile in the context of climate change. Policy and scientific research communities have suggested a number of technological approaches to mitigate these emissions. This paper uses the concept of co-production as an analytical framework to understand farmers' evaluation of a 'good animal'. It examines how technology and sheep and beef cattle are co-produced in the context of concerns about the climate change impact of methane. Drawing on 42 semi-structured interviews, this paper demonstrates that methane emissions are viewed as a natural and integral part of sheep and beef cattle by farmers, rather than as a pollutant. Sheep and beef cattle farmers in the UK are found to be an extremely heterogeneous group that need to be understood in their specific social, environmental and consumer contexts. Some are more amenable to appropriating methane reducing measures than others, but largely because animals are already co-constructed from the natural and the technical for reasons of increased production efficiency.


## KEYWORDS

Climate change Sheep Cattle Methane Genetics Co-production


## 1. INTRODUCTION

Sustainable intensification of agriculture (Foresight. The Future of Food and Farming 2011) has become one of the major policy and research themes in the UK. It aims to achieve increased food production while limiting impact on the wider environment and ecosystems. While the term 'sustainable intensification' is contested (Garnett & Godfray 2012), the approach has focussed attention on the importance of agriculture, as a source of food, a cause of environmental damage and provider of ecosystem services. In the context of animal agriculture, animal welfare is also an important factor. This complexity of objectives – providing food, ecosystem services and landscapes, while reducing pressure on the environment, and mitigating and adapting to climate change – presents a challenge to farmers and provides the canvas for this examination of animal-technology co-productions.

Greenhouse gas emissions from agriculture, particularly methane emissions from sheep and cattle have gained an increasing profile in the policy and academic worlds since the publication of a seminal report from the Food and Agriculture Organization of the United Nations (FAO) on the environmental impact of farm animals (Steinfeld *et al.* 2006). From a climate change point of view, methane and nitrous oxide are both important but this paper will focus only on methane. Methane from ruminant animals (e.g. sheep, cattle, goats, but not pigs) is a product from the digestion of grass, and is estimated to be around 25 times more powerful as a greenhouse gas than carbon dioxide (Industry Delivery Partners 2011). Put simply, in terms of the amount of greenhouse warming potential generated per tonne of meat produced, pig meat produces about 4 tonnes as compared to nearly 15 tonnes for beef and nearly 16 tonnes for sheep (Genesis Faraday Partnership 2008 based on analysis by Cranfield University). In the UK, a key climate change policy event was the passing of the UK Climate Change (2008) Act. This Act mandates a reduction in greenhouse gas emissions from the UK by 80% by 2050 (from 1990 levels). Interim targets for different sectors of the economy have been set by the UK Climate Change Committee, a body created by Government to advise on how this reduction can be achieved. Emissions attributed to the agricultural sector are estimated at around 8% of total UK greenhouse gas emissions, with 38% derived from methane (Committee on Climate Change 2010).

A range of responses have been suggested to achieve reductions in methane emissions from sheep and beef cattle. These include measures to increase the efficiency of production, for example by improvements to animal health, reproduction and genetics. Another approach is to manipulate methane emissions, for example by use of feed additives, using grasses which result in lower methane releases or by vaccinating the rumen to alter the bacterial composition. The agricultural industry has responded by developing a greenhouse gas action plan with targets and priorities (Industry Delivery Partners 2011). These include a wide range of actions, such as improved animal nutrition and health, the use of genetics and improved grassland management. The plan has been developed through a partnership, which includes a large number of agricultural organisations, such as the National Farmers Union, as well as more ecologically-focussed organisations, such as the Farming and Wildlife Advisory Group and Elm Farm Organic Research Centre.

(Moran *et al.* 2011) have calculated Marginal Abatement Cost Curves that highlight actions expected to give the best greenhouse gas response with least damaging economic impacts (or even in some cases positive impacts). They identify feed additives and genetic selection of beef cattle as measures with apparently good potential. Life cycle analyses developed by Cranfield University (Genesis Faraday Partnership 2008) highlighted the impact of genetic change, calculating that using genetics has reduced greenhouse gas emissions from pigs by 15% between 1988 and 2007. Beef cattle and sheep methane emissions were reduced by less than 1% during the same period, attributed to the slower adoption of genetic evaluation in these sectors. Jones *et al.* (2013) combine expert and farmer opinions and identify using legumes in grass mixtures as ranking highly as a way of mitigating greenhouse gas emissions in UK sheep production. Gill *et al.* (2010) advocate expressing efficiency in terms of production of 'human edible food', to acknowledge that ruminants convert a non-human edible food (grass) into a human edible food. Garnett has written extensively on the subject (e.g. Garnett 2011) highlighting the potential adverse environmental and ethical consequences that may arise from excessive focus on production efficiency as a way of achieving methane reductions. A number of authors have linked production to consumption patterns and advocated reduction in meat consumption in rich, Northern countries as a way of reducing greenhouse gas emissions (e.g. Garnett 2011).

While the entire food chain is important and is also influenced by consumption patterns, this paper will focus mainly on the farm end of the chain, touching on consumers and the wider food chain only in so far as they influence on-farm activity. The framework of co-production of the animal and the technical, will be used in this paper. A focus on co-production avoids the tendency of veering either too much into the realm of a technical fix, or of emphasising the social at the expense of the natural (Goodman 1999). Aspects relating to methane as a climate change gas will be foregrounded as the context for co-production. I will first locate this paper in the co-production and climate change literature and then examine how animal-technical co-productions narrativise the conceptualisation of methane and how methane is understood by farmers.

## 2. IDEAS OF CO-PRODUCTION

The concept of co-production in this paper is situated in animal geography (e.g. Holloway 2007). Co-production is also a concept used extensively in science policy studies (Jasanoff 2006) and innovation studies (e.g. Schot & Rip 1996), although with somewhat different emphases (Doubleday 2007). The science policy approach focuses on how categories of natural and social are brought into being; a process which involves the "intertwining of the cognitive, the material, the social and the normative" (Jasanoff 2006, p6). A different (but complementary) understanding of co-production comes from innovation theory. Here the co-production refers to the way in which technological change is brought about by merging technical innovation and social innovation–processes found to be almost an inevitable part of technological change (e.g. Schot and Rip 1996).This latter approach draws on the frequent finding that impacts from technological innovation do not just happen, but rather require the active involvement of a range of different stakeholders, including end-users, to bring them about. When considering sheep and beef cattle, co-production can be understood in both the above senses. Using the former approach, the emphasis is on different concepts of animal breeding and ways of evaluating animals. Using the latter approach, the focus is on social structures enabling the development of methods of estimating genetic value, the distribution of 'genetic material' (either as animals or in the form of semen), and so on.

The animal geography approach to co-production has been influenced by the perspective that technological change in agriculture is bringing about changes in human-farm animal relationships in ways that result in reconstituting the animals (Holloway & Bear 2011). Holloway (2007) gives a good example of this reconstitution from studying robotic milking of dairy cows. While the robotic milking system is intended to allow cows the freedom to act naturally and be milked to the cow's agenda, cows that do not adapt to the robotic milking system are culled. In this way the dairy cows are co-produced to fit the robotic milking system.

In comparison to farming of other species, sheep and beef cattle in the UK have been relatively little impacted on by technology. One of the main technologies applied is genetics. Holloway and colleagues have extensively examined the social construction of genetics and breeding value in the context of beef cattle. Holloway (2004) demonstrated how animals are evaluated by farmers in a show ring context, where aesthetics as well as utility become important. In contrast, at the crux of technological genetic evaluation are Estimated Breeding Values or EBVs, statistically determined quantifications of genetic value. However, the link between an EBV and the genetic composition of the animal is not necessarily taken for granted or accepted by farmers

(Holloway 2004). Even within the context of EBVs, genetic evaluations are not a single entity. Holloway *et al.* (2009) describe separate evaluations for meat production and calf production (Beef Value and Calving Value indexes, respectively), as well as the way in which individual animals are identified as contributing particular genetic qualities to the herd. Holloway & Morris (2008) demonstrate that animals consist not only of the material (the animal) but also the genetic information that goes with them. Holloway *et al.* 2011) examine how values attached to animals are produced within, and affect, knowledge practices and describe the entangling of 'traditional' and 'genetic' modes of valuation. Genetic variation can also be viewed more broadly as a symbol of the farm family's success (Convery *et al.* 2004).

As with Holloway's (2004) pedigree breeders, Calvert (2013) emphasises the importance of the concept of 'breed' and its social construction within cattle breeding practices in the USA. Grasseni (2007) perceptively observes the dichotomy between the need to create a 'breed' from individual animals of similar appearance, and selecting outliers from this group to produce the next generation, emphasising specific characteristics. Uniformity is first needed, in order then to identify distinctiveness. Breed can also be used to subvert technical expertise. Grasseni (2007) describes how dairy cattle breeders, in Italian mountain regions, deliberately undertook experimental crossing with a range of different breeds as a way of emphasising their disengagement and independence from agricultural and breeding advisors.

Once established, a breed identity must also be maintained. Holloway *et al.* (2009) examine the pivotal role of UK breed societies in delineating a breed, by controlling entry to the breed and recording relationships among individuals in the breed. They suggest that breed societies, by their actions, are also implicated in the co-production of both humans and farm animals. Calvert (2013) identifies similar practices involved in pedigree Aberdeen Angus breeding in a more industrialised context in the USA.

Beef cattle and (to a lesser extent) sheep genetic practices rely on more than local knowledge. Gibbs *et al.* (2009) note how specialist genetics knowledge can be located at a considerable distance from farms in corporate facilities, highlighting the digital and mobile nature of EBVs.

Once the 'body' of an animal has been converted into computerised 'data', it can be readily shared widely across computer networks.

Grasseni (2007) emphasises the management changes that become necessary as a result of adopting expert-led breeding programmes, for example, the resulting increase in size of cows leading to the need for better diets and increased stall sizes, and the need for infrastructure to supply semen. Thus, one technical change becomes predicated on a whole range of new practices and infrastructural accommodations in a form of animal-technical co-productions. In her specific case study of dairy cows in Italy, the adoption of standards and technical approaches also led to the need for professional specialisation between family members, and increased liaison with a wide-range of organisations such as vets and advisors on breeding and nutrition. The introduction of technical genetics, therefore, did not exist in isolation but required a reconceptualisation of the whole farming practice.

Beyond the co-construction with genetics, multiple and complex relationships have been found between humans and farm animals, both within the context of farmer relationships with farm animals (Riley 2011, Wilkie 2005, Convery *et al.* 2004) as well as the relationships with farm animals of a range of different publics (Macnaghten 2004). These emphasise the extension of relationships beyond a merely economic, production orientation.

Ruminants, as grass-eating animals, are often associated with particular environments. Ruminants can be considered important for maintaining biodiversity and identified with a sense of the local (Evans *et al.* 2003; Yarwood & Evans , Yarwood & Evans Yarwood & Evans2003). Yarwood & Evans (2006) suggest co-production of animal and locality takes place at the breed level. Gray (1998) goes further and suggests that for hill sheep farmers, knowledge of local sheep becomes essential in the breeding of specific flocks for specific places, with sheep becoming adapted or 'hefted' to specific areas of land within a region. This type of conceptualisation is not necessarily applicable across national boundaries, for example, according to Saltzman *et al.* (2011) cattle are perceived to belong to Swedish nature, but as foreign to Australian nature.

Haggerty *et al*. (2009) stress the dynamic relationship among landscapes, animal bodies and climate in a New Zealand sheep farming context. These relationships in turn are sites of contestation of what constitutes 'good' farming (Burton et al. 2008) and have influenced the development and shape of sheep farming. As a result, Haggerty *et al.* (2009) argue that New Zealand sheep farms have developed in two different directions. Positive re-enforcement between productivism and the accrual of economic and cultural capital led some sheep farmers in a productivist direction, where improving productivity implied improving the animal body. Yet others responded to environmental and animal health challenges that began to arise by moving to new market linkages, with alternative conceptions of what is the 'good' farmer. This bifurcation is also seen elsewhere. Grasseni (2007), for example, argues that breeding technologies are used by marginal farmers either as a means of converting to intensive farming or to pursue their identity as champions of both genetic and cultural capital.

This emergence of alternative conceptions of the 'good' farmer engenders different practices of co-production with more emphasis on the consumer. Stassart & Whatmore (2003) provide a detailed analysis of the development of new alignments between consumers and producers, in the context of Belgian beef production. Stressing the ontological continuity between human and animal bodies, and responding to anxieties around industrial meat production, they suggest that meat is a co-production of the social and the natural, where healthy food is predicated on healthy animals. However, this co-production requires the complicity of a number of intermediaries, including butchers and certification bodies.

Co-productions have been identified between animal and technical (e.g. Holloway 2007), animal and production systems (e.g. Haggerty *et al.*2009), animal and environment (e.g. Yarwood & Evans 2006) and animal and consumer (Stassart & Whatmore 2003). This paper seeks to extend this range of examinations to consider inter-relationships with technologies to reduce methane emissions. I will firstly examine the literature relating farmers to climate change and to pollution more generally.

# 3. FARMERS, POLLUTION AND CLIMATE CHANGE

Greenhouse gases can be considered a form of pollution. The seminal work of Douglas (1966) emphasised the moral nature of pollution. Pollution, she found, was culturally identified as something to be disapproved of; as 'matter out of place'. This moral aspect to pollution in a farming context was echoed by Lowe and colleagues (Lowe *et al.* 1997) studying water pollution from dairy farming. They identified two possible ways in which pollution could be framed; either as an unavoidable side effect of valuable production activities or as a morally discreditable practice, something that attracts blame. They explain "When defending themselves against charges of damaging the environment we found that farmers were able to draw on a rich repertoire of justification, which essentially naturalized their actions" (p192).

From this same empirical work, Ward *et al.* (1998) reported how dairy farmers tended to modify their practices to reduce water pollution in response to regulatory pressures, but in a rather "begrudging" (p1172) way partly because farmers perceived criticism as expression of a wider 'anti-farmer' attitude in society rather than valid comment on sources of pollution. Ward *et al.* suggested that farmers responded in an 'add-on' and 'end-of-pipe' way rather than seriously engaging with the sources of pollution. Similarly, Ward *et al.* (1995; 1207) reported that "the group felt that agricultural pollution was far less of a problem than industrial pollution and suspected that farmers were being more strictly regulated because they were 'easy targets'.

This refusal to accept the 'pollution' framing for farming activities is not restricted to the dairy farms studied by Ward *et al.* (1998). Diffuse water pollution from agriculture, in particular, is often not visible and its impacts are also not visible. Therefore, farmers tend not to be convinced that there is a problem or that they are part of the problem and can do something about it (Blackstock *et al.* 2010).

Climate change can also be viewed in terms of a moral issue, but it is more often conceived of in terms of risk. Survey evidence suggests wider publics are concerned about climate change and perceive the effects of climate change to pose a potentially serious risk, but this risk is expected to impact on other people. The risks to individuals themselves are perceived as being low (Spence *et al.* 2010, Spence & Pidgeon2010). Furthermore, climate change may even be perceived as having beneficial effects locally, e.g. due to warmer weather (Spence & Pidgeon 2010). Little research evidence exists specifically on farmer approaches to climate change. However, in common with wider publics, Holloway & Ilbery (1996) found UK arable farmers identified both beneficial and harmful effects of global warming. Holloway (1999) noted how arable farmers stressed the long timescales involved and difficulty of planning for unpredictable local climatic changes, as well as a certain amount of scepticism regarding the existence of climate change. Barnes and Toma (2012) surveyed 540 dairy farmers in Scotland and found only 48% agreed that temperatures would rise in the future. Survey data from Californian farmers (Haden *et al.* 2012) suggests that farmers align with wider publics in perceiving climate change impact as being something primarily affecting other people.

Kahneman & Tversky (1982) identified that people generally will take risks to avoid losses, but are more averse to risking what they already have in order to gain more. Spence & Pidgeon (2010) apply these findings to attitudes to climate change and suggest that since wider publics do not perceive climate change as a personal threat, they are less likely to wish to risk what they have at the moment, in order to take action to mitigate climate change. Haden *et al.* (2012) extend this approach to farmers, and identify that Californian farmers were more likely to undertake practices to adapt to climate change since this is seen as a personal threat, rather than adopt mitigation practices which rely on a wider sense of doing good for others. Therefore, Haden *et al.* (2012) suggest that farmers are more likely to adopt mitigation practices which also have a tangible benefit to the individuals involved (e.g. nitrogen management, using less energy) than those of more general impact.

Actions of farmers can also be influenced by consumer pressure. Consumers in the UK increasingly associate perceptions of food quality with perceptions of the natural (e.g. Murdoch *et al.* 2000) linking into ideas about the benefits of local production, embedded in local geographies. Eden *et al.* (2008) established that farmers markets are valued, in part because they are perceived to reconnect the consumer and the producer. Features such as dirt on vegetables, was used as evidence of local, authentic production. Gilg *et al.* (2005) argue that a shift to sustainable consumption is likely to be a component of a wider shift to sustainable lifestyles, but this shift is predicated on feeling that actions taken will be effective and have a positive impact on the environment. However, Barr *et al.* (2011) found that climate change mitigation was unimportant in the context of sustainable consumption, noting that 'climate change is for all intents and purposes, a relatively minor issue when it comes to embedding environmental practice' (p 3025). This evidence suggests that consumers will not readily identify food consumption with climate change mitigation activity. Most consumption, however, is the end result of a long and often complex food chain. Green & Foster (2005) therefore argue that sustainable food production requires consideration of more than production and needs to include socio-technological practices across the entire food production system.

The research reported here extends the research on water pollution, climate change and consumer links with producers to consider methane emissions from sheep and beef cattle. I examine how farmers conceptualised methane, its role in climate change and potential mitigation methods. The research contributes to the body of knowledge around co-production by examining concepts of animals in a novel context and in the face of a novel challenge.

# 4. RESEARCH METHOD

The research sought to elucidate how individuals understand their world and their context and a qualitative methodology was adopted. I conducted 42 semi-structured interviews with farmers and other actors in the farming sector between September 2010 and March 2011. To reduce potential bias, a semi-structured interviewing approach provided a basic structure to the interviews and allowed greater flexibility to deeper explore specific items. Questions were focussed around how breeding replacements were chosen, how markets for breeding animals and meat influenced the breeding decisions, the perceived importance of environmental issues, and more specifically, the perceived importance of methane emissions as a factor to be considered. Interview schedules were informed by four preliminary, open-ended, scoping interviews with a range of industry representatives, including a representative of the meat industry, a commercial advisor to farmers, a specialist in native breeds and a provider of social support to farmers.

The 42 respondents included 30 sheep and beef cattle farmers and 12 representatives from the wider industry (vets, consultants, meat processors, breeding advisors, researchers). In order to avoid interviewing only elite farmers, interviewees were identified through a range of different 'gate-keepers' including snowballing from preliminary and main interviews, recruitment through veterinary practices and farmers living near existing contacts in rural areas. Interviewees were purposively sampled to provide diversity and avoid bias as far as possible (Ritchie *et al.* 2003). To this end, particular efforts were made to ensure representation from organic and environmentally-oriented farmers. Farms were located in a diverse range of environments, including upland and lowland, moor and marsh. A specific strategy was to target several different geographical areas within England and Scotland as a way of achieving variation in types of farming enterprise.

Farmers were interviewed from a wide-range of production systems, sizes and producing for different markets. Markets included producing breeding animals for direct sale or sale by auction; sale of meat animals either to other farmers for growing-on ('store' animals), or as animals for slaughter; through auction, direct to meat processors , for specific supermarkets, or meat for direct sale in farm shops, to hotels or by post.

On approach for interview, farmers were given an information sheet outlining the purpose of the interview, including its focus on methane. The interview situation was not treated as a data mining exercise but rather that meaning was to be created in the interview encounter (Holstein & Gubrium 2002).

Interviewees included specialist sheep and beef farmers, but many farmers kept both species. Four dairy herds were included as producers of beef from beef-cross-dairy animals. Farms ranged in size from 4–1,000 beef cows and 8 – 1,300 ewes. Farmers were invited to complete a questionnaire to provide further information on the interview sample. 23 farmers completed this questionnaire. From these, it transpired that 30% of the farms self-identified as hill farms, 52% as not hill farms, with the remainder considering themselves marginal or upland farms. The farms relied heavily on animals for their income with 70% indicating that more than 50% of their finances came from animals. The majority of farms (91%) employed up to 5 people and most (83%) reported frequently using computers. Only 9% reported rarely using computers, and in at least one case this was due to the unavailability of broadband in the area. 43% of the farmers were aged 35–54, with 17% aged over 65 and 9% less than 35. The farmers were very experienced, 87% reporting more than 10 years of farming experience and 57% more than 30 years. On the basis of interview information, around 50% of farmers made use of technical genetic evaluations in the form of Estimated Breeding Values for beef, but less than 20% for sheep. However, this statistic is difficult to interpret as the extent of reliance on the Estimated Breeding Values varied. Some farmers stated they were primarily interested in the appearance of the animal but would also look at the Estimated Breeding Value.

Interviews with farmers were all conducted by the author and primarily face-to-face on farm, although a small number were undertaken as telephone interviews for logistical reasons. Most other respondents were interviewed in their place of work or by telephone. Interviews were recorded (with permission), transcribed and then analysed inductively, seeking to identify themes across the interviews. The interview data are not presented as statistically valid samples, but rather as a rigorous engagement with as wide a range of views as possible. Quotes used in this paper are illustrative of themes identified in the transcripts. Interviewees are identified by number and farming-type. Preliminary findings were presented to a one-day stakeholder workshop attended by 10 stakeholders and 2 academics. Feedback from this workshop was taken into account in finalising the analysis.

This qualitative approach has enabled understanding of how farmers view technological approaches to mitigating methane in their specific situations and to drawn inferences on how animals are conceptualised in a range of different circumstances. The data collection method does not allow for in-depth study of co-production in one particular context but is powerful in drawing on a large range of different contexts for a meta-level analysis.

# 5. RESEARCH FINDINGS

In analysing the data, I have endeavoured to link farmer knowledge about animals with technical knowledge about animals. In doing this, I have paid attention to the different ways of knowing and the different objectives of the farmers. Furthermore, grazing animals always exist in specific natural and consumer environments, which impact on how farm animals are conceptualised.

The policy imperative to reduce methane emissions has tended to focus on solutions that depend on improving efficiency of production in terms of the amount of methane produced per unit of meat produced. Efficiency measures include a range of practices such as use of genetics to increase growth rate and efficiency of feed use, and improved reproductive performance and health. Beyond these a range of specific measures to reduce methane have been suggested, notably different varieties of grass and feed additive to reduce methane emissions, and alteration of rumen bacterial populations by vaccination to reduce methane emissions. Some farmers embrace this emphasis on efficiency and respond by adopting a range of efficiency enhancing measures and innovations. However, the motivation for such actions is driven more by economic considerations rather than methane reduction.

Cattle and sheep were constructed by farmers as means of providing a livelihood from a particular physical location. These were often in harsh environments particularly where sheep were concerned. In these situations, famers tended to emphasise the coproduction of their animals with the environment, such that their animals were particularly suited to the environmental niche they found themselves in. Manipulation of the animals, and to a large extent, animal husbandry methods, was considered risky, and therefore resisted.

Animals were also co-constructed as meat, particularly by farmers who sold meat products direct to consumers. For these farmers, demands of consumers and consumer perceptions of good quality meat became critical in determining desirable characteristics of their animals. These desirable characteristics were often in opposition to measures required to reduce methane emissions.

In the next sections, I will explore in more depth these three types of production, namely: focus on production efficiency, focus on production in demanding environments and focus on production for consumers. Each of these contexts draws out different aspects of animal-technical co-productions. By selecting themes that cross a range of farming approaches, I draw some generic principles that are likely to apply more widely across farms. While individual farms may emphasise one particular characteristic (like productivity) my broad-scale approach allows farms to appear in multiple categories, reflecting the complex ways in which individual farms are addressing the challenges of producing food, while minimising ecological impact and reducing greenhouse gas emissions. Finally, I will consider how farmers make sense of methane, demonstrating how this is conceptualised as an integral and inevitable part of an animal that cannot be disaggregated from the animal, rather than as a pollutant, by-product of beef and sheep-meat production. The contrast between the policy imperative to reduce methane and the farmers constructions of their animals are then highlighted in a final section.

# 6. ANIMALS AND EFFICIENCY OF PRODUCTION

Climate change policy (and much of the associated natural science research) agenda is predicated around a description of the methane challenge as one that is technocratic in nature. Such a framing emphasises the ability of technology to reduce the amount of methane produced per animal and hence reduce climate change impact, in large part by increasing the efficiency of production. The implication is that the material animal can be disaggregated (cognitively) into its component parts and these parts manipulated to create a new animal that is more desirable from a climate change point of view.

For farmers with a predominantly productivist approach to agriculture, the application of technology was perceived as a normal part of efficient farming. Emphasis was placed on the animal as a production unit and prominence was given to terms used in accountancy, such as gross margins and efficiency of resource use, as keys to evaluation of a 'good' farmer.

Individual breeds were highlighted, but cross-breeding was also practiced and hybrid or 'synthetic' breeds specifically developed. Market competitiveness was emphasised as evidenced in the following quote: "*There will be a lot of breeders select purely on looks of an animal. Our breeding aim of our breed is to try and improve the profitability of our customers, which is completely different to any other pedigree breeder in the country*" (interview 102[a], sheep breeder). Similarly EBVs were embraced, for example, "*I want to know how that animal is going to perform, what it's going to give me on my farm, so EBVs just seemed a very natural way of doing that really*" (interview 99, beef farmer). The focus was on performance measures rather than appearance, and economic performance rather than aesthetics. Any technologies advocated to reduce methane that also improved efficiency of production were seen as acceptable. New grass varieties producing less methane and new feed additives to reduced methane production would be acceptable as long as economic efficiency was unaffected or improved. Use of vaccines to alter rumen microbial populations to reduce methane emissions could be accepted subject to impact on overall efficiency and financial impact. Genetic techniques and disease reducing measures were already incorporated into farming practices.

The animal was perceived as an efficient converter of grass to meat (although often with additional cereal diets provided) and the emphasis was on, for example, residual feed intake (how much feed is needed beyond that strictly necessary for growth). Other technologies to increase efficiency were adopted as appropriate, e.g. automatic oestrus detecting equipment in order to increase the number of cows achieving pregnancy quickly and enabling use of high genetic merit sires. Unlike dairy cattle, beef cattle are not frequently observed by humans, therefore oestrus detection tends to be left to a bull. However, using a bull rather than artificial insemination (AI) limits the ability to use sires evaluated as having high genetic merit.

One prominent trend was towards minimising human interventions and hence improving economic efficiency. As an example, one strand of sheep breeding focuses on selecting animals that are robust, disease resistant and requiring minimal human assistance (e.g. with lambing), and even shedding its own wool (the cost of shearing exceeding the value of the wool gathered at the time of interviewing). While the methods used in this approach are very much technocratic, emphasising selection on the basis of numerical recording; the target traits (lambing, diseases, wool shedding) emphasise a return to the natural. As with the use of robotic milking (Holloway 2007), technology is used to reduce the amount of human intervention. However, in the case of introducing wool-shedding, a more radical physical change is introduced to the animal body, although the wool-shedding characteristic is introduced through a wool-shedding gene that occurs naturally in some breeds of sheep. The sheep is, then, a co-production of the material, physical characteristics that enable this modification to the sheep body, cognitive processes that value the 'natural', but with a social emphasis on economics of producing sheep for meat, the high cost of labour and normative evaluations that recognise the appropriateness of using technology to achieve these aims.

Although the climate change policy and research focus has been largely on beef produced from beef cattle, in the UK around 50% of beef is supplied from dairy herds, much of it as beef-cross-dairy calves used to produce the pregnancies that ensure milk production. In terms of formal methane accounting, emissions are often completely attributed to the dairy cow. Nevertheless, beef-cross calves will produce methane, and are therefore, also an appropriate target for methane reducing measures. Among the dairy farmers interviewed core activity was seen as producing milk, with the beef-cross calf a 'by-product' or more positively 'icing-on-the-cake' and calves sold on to others for rearing as quickly as possible. The emphasis was on the cow as a reproductive unit, and AI semen was selected on the basis of good fertility and ease of calving.

Beef-cross calves were conceptualised by dairy farmers largely by their impact on the dairy cow, rather than as themselves. This construction of the calf is very different to that of farmers with beef production as part of their core objectives. For dairy farmers there is a clear distinction between 'valuable' (dairy) calves and 'disposable' (beef-cross and dairy bull calves) produced largely due to the specialisation inherent in current dairy farming practices. Manipulating the efficiency of the beef calf by selecting sires with genetic potential for fast, efficient growth was of little or no concern. In this way the calf is co-produced specifically within a type of production system. The calf is a crucial part of specialist beef production but a by-product of specialist milk production. The disjunction between producing milk and producing beef has led to calls for considering a return to dual-purpose cattle where both dairy and beef are produced from the same breed of animal, in order to reduce methane emissions. However, interviewees highlighted the infrastructural and management demands that a return to dual-purpose cattle would make.

While the impact of methane produced by sheep and beef cattle was rejected by the farmers interviewed, compliance with policy pressure, and synergies between production-oriented farming and technological measures to reduce methane meant that, at minimum, the rhetoric of methane reduction was accepted and appropriate mitigation practices adopted by some farmers.

As noted earlier, sheep and beef cattle production in the UK tends to take place in more marginal agricultural areas where environmental challenges are emphasised. It is to these that I turn next.

## 7. ANIMALS IN ENVIRONMENTS

Despite the focus on production and animal bodies, agriculture, particularly ruminant agriculture, can rarely escape the influence of landscapes and climates (Haggerty *et al.* 2009). As production moves to more marginal land, landscape and climate come to dominate. In stark contrast, one of the first tasks for a geneticist is to use statistical methods to separate out the genetic from the environmental, in order to understand and manipulate the genetic (with acknowledgement of possible interactions between environment and genetics). Inevitably, therefore, landscapes and climates contribute to the construction of the animal. From the farmers interviewed, three strong themes emerged, although sometimes more than one was held at the same time. The first theme emphasised the physical challenges presented by environments, the second the use of sheep and beef cattle for managing biodiversity, and the third as the complex inter-related nature of grazing animals, humans and other aspects of environments.

The restrictions placed by working in challenging environments were particularly emphasised by farmers in upland areas, where the vagaries of the weather and topography strongly influenced activities. Animals were identified as being particularly adapted to a specific environment with sheep 'hefted' to a particular piece of land (see also Gray 1998) and managed in ways that allowed them to thrive on that land. As a result of the emphasis on the harshness of farm environment, the materiality of the animals (particularly sheep) was strongly embedded with specific locations. Farmers stressed physical aspects, such as having a robust frame to be able to cope with strong winds. Visual assessment was emphasised, as well as handling the animal in order to evaluate its condition, rather than relying on figures. There was a degree of co-operation between the farmer and the animal. Farmers asserted that sheep learn how to thrive in a specific environment from each other, but farmers assisted in animals' adaptation to unfamiliar circumstances. For example, one farmer described how he was now teaching his sheep to eat hay, so that they would be able to rely on hay that he provided during snowy weather. During the previous snowy winter, his sheep had been unfamiliar with the hay provided for them and unwilling to eat it, even when there was no other food available.

EBVs were often viewed as appropriate for more intensive, lowland conditions without strong environmental constraints. Upland farmers emphasised how their stock was adapted to their environments, for example, although growing slowly, they would continue to grow over a longer period of time and achieve better weights than incoming animals. Incoming animals from elsewhere struggled to thrive as described by this interviewee: "*being ... a very marginal hill farm...If we buy in [animals] they need a year to acclimatise anyway, so we're far better off to keep our own [replacement animals] if we can*" (interview 116, hill sheep and beef farmer). Farmers talked about trying new breeds or new sources of their existing breeds with apparently desirable characteristics, but which proved unsuitable on their farms. For farmers in marginal areas, the challenging environments led to an emphasis on the ability of the animal to survive and remain productive in these conditions. Sheep (and to a lesser extent cattle) were considered integral to the environment, although also challenged by it.

Not only was the ability of the animals (and the farmer) to survive on hill land important for their identity, but farmer livelihood in many cases relied on payments for environmental management, particularly biodiversity conservation through various agri-environmental schemes. These schemes implied restrictions on the numbers of animals grazed, timing of grazing, and in a few cases, restrictions on providing animals with additional feed. For some, environmental management considerations were cited as reducing their freedom to change practices, e.g. *"I'm not sure what we can do to reduce it [methane], less cattle maybe but then we graze a large area and we're driven by the number of cattle it takes to graze those marshes to a standard rather than fill in the marshes with too many cows"* (interview 95, beef farmer). These restrictions prevented farmers from considering re-seeding with less methane producing grasses, or sometimes even providing feed additives to reduce methane emissions. The animal became strongly identified with managing biodiversity.

At the extreme, biodiversity management became the main reason for keeping animals, as exemplified by one interviewee undertaking conservation grazing for an environmental body. Here again, the ability of the animals to thrive on rough pasture was emphasised, as was the ease of handling and management, for example, *"from my point of view, having animals around for longer is actually a real benefit because I observe that as they get older they become better adapted at this kind of system and most of my management problems are focused on the younger animals"* (interview 103, conservation grazer). Rapid growth is not necessary to the definition of a 'good' animal in this context.

The emphasis on biodiversity management did not necessarily imply a co-construction of the animal and the environment. For example, an approach that emphasised efficient production could be combined with management to achieve biodiversity aims by altering the number of animals grazing rather than impacting on the animals themselves. The animal itself remains unaffected by the aspiration to manage biodiversity; rather it becomes a tool in the process. However, in other cases, such as where native breeds were selected to achieve biodiversity management, biodiversity considerations became part of the construction of the animal.

A complementary viewpoint emphasised the wider, systemic aspect of their farming approach, and the interlocked nature of the different aspects of the farming system. Animals were described as a part of complex systems which included grass, soil and trees, as exemplified by this hill farmer: "*you're impacting one way or another on the environment in which you live, and how do you work with that, not to be against it but to work with it? ... we once went down the road of having a much larger number of sheep here and I just felt, a) I was stressed, b) they were stressed c) I was overgrazing... So...we relaxed, or I did and said, wait a minute, we've all got to be part of this, so whether it's a sheep or whether it's a cow or whether it's a deer or a hare on the hill, it seems to me the whole thing is completely interlocked*" (interview 105, hill sheep farmer).

The emphasis on farm animals as part of wider, complex systems led to an emphasis on managing methane emissions throughout these systems. Reductionist, technical approaches were viewed as inappropriate, for example "*Why breed for animals that are producing less methane when actually you could address the issues with management. In a way, producing animals that produce less methane is ignoring or avoiding the actual land management methods that can be used to control, or at least manage carbon*" (interview 112 , organic mixed farmer).

Animals were perceived as a means of providing a livelihood from a particular physical location. The physical animal was required to produce beef or sheep meat mainly or entirely from grass and to be robust to the physical conditions prevailing in the specific location. In this way the location co-produced the animal. Biodiversity could be viewed as another product that did not impact back on the animal. Rather the animal produced the biodiversity. However, animals could also be evaluated on the basis of their ability to produce biodiversity or the ease of management in circumstances where contact with humans was limited. Environmental impacts on the animal are, therefore, various, and other more complex interactions can exist, in part influenced by specific locations and farmer values.

# 8. ANIMALS AS FOOD

Farmers held two contrasting views on the 'good' animal for food, depending on whether animals were sold to intermediaries or directly to consumers.

For those farmers who supplied meat direct to consumers, perceptions of consumer demand were fundamental to defining the animal. The consumer was perceived to demand natural, local and good tasting meat.

Naturalness is associated with grass fed animals. Thus, these farmers emphasised producing sheep and beef cattle from grass, in so far as was possible. Traditional, native breeds were preferred as they were perceived to be able to 'finish' from grass (achieve an appropriate level of fat and muscle, as opposed to mere size). This preference meant that animals were bred for systems, rather than manipulating the environment to suit animals bred under different circumstances. These native breeds were perceived as able to perform well in UK conditions, unlike the 'continental' breeds more commonly used in beef production. Less emphasis was given to growth rate, as slower growth rate was perceived as associated with better meat quality. Priority was also given to continuity of supply throughout the year (for beef), despite grass-based production systems being seasonal in their nature. Slower-growing animals provided continuing supply and were, therefore, valued as much as faster growing ones. There was no imperative to slaughter animals as quickly as possible. In constructing the animal for the direct-sales market, naturalness and continuity of supply were premium concerns.

For these farmers, the animal is co-constituted not just within its environment but in terms of how it can best meet high quality consumer desires. The cow or sheep is a link between the land and the consumer, while also remaining a part of the land, not apart from it. The technical may or may not be seen as part of this locus.

The emphasis on the natural had impacts on aspects of farm management, for example feed additives were avoided and vaccines only used when it was absolutely essential, for example, "*I'm not ill-disposed to a feed additive. What we are very careful about is just what we put in feed additives here because one of our features is that we put as little as possible into the cattle. At one stage there was also feed additives and probiotics were the feature of the month and that's not what we want to come with our customers because when they hear feed additives they tend to go aagh, irrespective of what's in them*" (interview 110, hill beef farmer).

An alternative view was expressed by farmers selling to intermediaries and therefore at a further distance from the consumer. These farmers were still interested in consumer perceptions, but consumer perceptions were mediated by, and often delegated to, intermediary bodies. Most farmers in this category were selling either directly to abattoirs/meat processors or selling animals at auction for purchase by meat processors. Here, the price received for the animals was a key focus of attention. Prices varied by season, demand and supply characteristics, and by weight and carcass grade. A standardised grading scheme is used for evaluating carcasses (known as the EUROP system), which is influenced by weight, visual evaluation of the fat cover on an animal and its physical shape (or conformation). The shape and fat cover on animals, therefore, become important determinants of the value of the animal to the farmer. This is reflected through to the animal breeders, who place considerable emphasis on animal shape.

In a further complication, farmers in more marginal lands are often unable to grow their animals to weights suitable for slaughter due to the unavailability of grass during the winter. Their animals are therefore sold on at lighter weights to other farmers, who grow the animals to slaughter weights. This is usually achieved through auctions of live animals during the autumn. As little information on the animals is available to purchasers, visual appraisal is predominant. Thus, the production of the 'good' animal for slaughter has a strong emphasis on shape, weight and amount of fat cover. All these features can be visually evaluated on farm. There is therefore a strong emphasis on the materiality of the animals, but in this case emphasising its imputed 'deadstock' value.

## 9. MAKING SENSE OF METHANE

Methane was perceived by the farmers interviewed as "*an inevitable consequence of keeping ruminants*" (interview 87, sheep breeder). From this perspective, as long as grass-eating, ruminating animals exist, then methane will continue to be produced. Measures to reduce methane were treated with caution with respect to the integrity of the animals as a whole, for example, "you *can't just take methane [and] isolate it*" (interview 26, hill beef and sheep farmer). Methane was viewed as also associated with wild animals, such as deer, that would naturally inhabit the environment currently occupied by sheep and cattle. Methane was thus construed as an integral part of sheep and beef cattle and in this way, methane also becomes part of the natural environment.

Methane does not have any of the features farmers associated with a pollutant. It is not synthetically developed by humans, nor is it associated with industrial activity or combustion of hydrocarbons. When thinking about greenhouse gas emissions, farmers would refer to cars, flying, industrial pollution and in some cases intensively reared feedlot beef cattle, associated with North American production systems, for example "*the bad press was all to do with feed lot animals in California or somewhere ...and that isn't quite the same as grazing animals making use of natural vegetation*" (interview 90c, sheep breeder). Thus 'excessive' methane is associated with perceived unnatural production methods, such as feedlot cattle systems. 'Acceptable' methane is associated with extensive production systems that are perceived to be more 'natural'.

The image of naturalness was largely associated with the fact that sheep and (to a lesser extent) beef, unlike poultry and pigs, are predominantly fed on grass rather than cereal-based diets, for example, "*my sheep... they don't get fed anything that makes them more gaseous, they're just eating what's naturally here ..So they're just doing what they've always done in a natural environment*" (interview 105, hill sheep farmer). As sheep and beef cattle eat mostly grass, and the grass absorbs carbon as it grows, farmers perceived that not only are there no external inputs that would cause pollution, but that growing grass has beneficial impacts. Sheep and beef cattle farming, in their view, would be expected to reduce greenhouse gas emissions due to the carbon captured in grass, or at minimum to be carbon neutral, rather than a source of greenhouse gases. Thus, arguments about the large climate change impact of sheep and beef cattle were not felt to be credible.

Extensive production systems were associated with traditional cultural practices and hence represent a human-animal association that has existed over historical periods of time. Since methane from these production systems has not caused a problem in the past, there was no expectation that it will cause problems in the future. Reference was made to the 'wisdom' of nature that has ensured that traditional sheep and beef cattle farming is in harmony with other factors in nature, as exemplified in this quote "*The whole world has been covered in ruminants from time began... Mother Nature, whatever that is, wouldn't have created a system that was going to destroy itself*" (interview 98, beef breeder). Rather than being a modern invention, sheep and beef cattle are part of an historic, natural order.

In the rhetoric of the farmers interviewed, sheep and beef cattle are not only a natural feature of the landscape but also have an important role in producing food for humans, managing the environment and providing a source of income, particularly in hill regions where there are perceived to be few other options for making a living, as exemplified in the following quote:

"*You want to go up to where [my wife] comes from and see if you can grow crops up there, good luck ... Steepness of slope, soil type, height above sea level, all those sorts of things just means there are vast areas that you can't grow crops on, so we have to utilise the grass that we can grow*" (interview 30, dairy and sheep farmer).

These arguments enlist sheep and beef cattle farming into the sphere of the moral 'good'. Farmers drew on repertoires of the intrinsic positive character of their management of nature in contrast to unnatural practices, as evidenced in the following quote:

*"I can't see what's wrong with what we're doing, we're producing a product that we can eat from a hill that otherwise would be either forestry, originally it would be forestry... I think we're doing a natural rearing method and I don't personally feel wrong about doing it, I feel confident that it's a good and right thing to do... I think there's plenty of other things in modern trappings that are doing more harm than that"* (interview 117, hill sheep farmer).

The evidence from these interviews suggests that there is little acceptance of methane emissions from sheep and cattle as a 'problem' or 'pollutant' requiring action by farmers. This message was remarkably consistent across all the production systems studied, including farmers with an inherent ideology of environmental care, such as some organic farmers. Methane was understood as an integral part of sheep and beef cattle and a necessary part of their grass-eating identity. Moreover, methane was seen as part of the specific farming context, particularly in the uplands, resonating with the culture of 'natural' production and the historical and cultural practices associated with these farming systems. Thus, to subvert Mary Douglas's (1966) famous description of pollution as 'matter out of place', for sheep and beef cattle farmers in the UK, methane is very much 'matter in place'. Sheep and beef cattle were seen as having a natural role in relation to humans, and methane as an inevitable by-product of this role. In the words of one hill sheep farmer, sheep (and by inference cattle) should be *"allowed their place in society"* (interview 21, sheep and beef farmer). In defining the grazing animal, methane production is also defined.

Methane was primarily conceptualised as an inevitable and natural part of sheep and beef cattle. As these animals contribute to the good of humankind by converting grass to a product edible by humans, there is a normative justification to allowing sheep and beef cattle to continue to exist. From a social perspective, farmers felt methane emissions were yet another criticism to be heaped onto beleaguered farmers who had more pressing issues to deal with. Arguments for the climate change impact of methane based on codified knowledge and scientific best estimates was perceived to be in conflict with knowledge that views methane emissions in the context of carbon cycles that take into account absorption by grass and the lack of introduction of 'artificial' factors.

## 10. CONTEXTUALISING CO-PRODUCTIONS

It is perhaps trite, but nevertheless important, to note that not all farmers or farms are the same. This is salient as the environmental context, market and ethos of the farm provide the location within which both the animal and the technical are constituted, and which therefore impact upon the way constructions are made.

Co-productions do not take place in isolated, sterile circumstances but are redolent with history of interactions among animals, farmers and technical specialists (Reardon 2001). In the case of the farmers interviewed here, a strong sense of historic injustice was communicated. Farmers felt they had adopted measures on advice from technical specialists, but these measured turned out to be disadvantageous, for example, "*Agriculture has always been pretty quick to take on board new things that were deemed to be good, but we've had quite a lot of stuff come at us that's deemed to bad, well it's turned out to be bad*" (interview 26, hill beef and sheep farmer). Farmers articulated examples from their experiences of BSE, Foot and Mouth Disease and environmental management - measures they had been asked to put in place and were now being asked to remove. They mentioned perceived bad genetics advice given to the dairy industry, resulting in lame and infertile cows due to excessive emphasis on milk yield. These experiences led to farmers' scepticism regarding technical expertise and reluctance to appropriate such knowledge in their practices.

The current framing of methane from farm animals assumes that methane is primarily the producers' rather than the consumers' concern. In purchasing white goods or cars, an energy rating is provided on appliances or efficiency figures on cars, providing the consumer with information that can enable choice. Yet in purchasing meat, choices are largely restricted to choosing the species (e.g. buying lamb or chicken) or responding to exhortations to reduce meat consumption. There seems to be little link between 'reduced methane' and consumers of sheep meat and beef. The benefits of reducing methane are expected to be gained purely from improved production and hence expected improved profitability for farmers. If the evidence of Barr and colleagues holds (Barr *et al.* 2011), consumers are taking little responsibility for climate change through their meat purchases and, therefore, are unlikely to demand low carbon footprint meat. However, consumer purchasing responses could readily change under different conditions such as pricing, availability, information, social pressures and so on. None of which are explored here.

The focus in the policy world on measuring quantity of greenhouse gas emissions means that the method of accounting for greenhouse gas emissions becomes crucial to understanding where policy levers can best be applied. Current practice is to use relatively crude calculations that impute a standard amount of methane emissions to each animal. Effectively it is the number of ruminant animals that are being measured, rather than methane emissions *per se*. For example, reductions of greenhouse gas emissions of 21% between 1990 and 2008 (Committee on Climate Change 2010) are explained as primarily due to a drop in the total number of animals, largely due to changes in the support regime that replaced payments for each animal with a Single Farm Payment (Gill *et al.* 2010). So, while the rhetoric of the policy world is to reduce methane emissions, the practice is to reduce the number of animals. At this point, there is a synergy between farmer and policy concepts, each understanding animals as a methane producing entity. While farmers struggle to view methane as being amenable to technological manipulation, policy makers look to more discriminating measures of methane that could take the impacts of technological manipulation into account.

From a scientific perspective, reflected in the views of non-farming interviewees, methane emissions are readily manipulable. Methods have been identified that have the potential to reduce emissions, with the added benefit of increased efficiency of output. In contrast, as described above, sheep and beef cattle are located in specific natural environments, production systems and markets and subject to appropriate practices and norms. Introduction of methane reduction as a new norm requires the introduction of a new form of social order that recognises low methane production as part of the constitution of a 'good' animal. In productivist approaches, this is already achieved through attention paid to efficiency of production, but in other approaches, particularly those which give high value to concepts of the 'natural', the association of methane reduction with intensification is anathema. The implication on evaluating the 'good' animal is that methane will continue to be constructed as a natural and inevitable part of the animal which is not a target for technological manipulation. If methane reductions are to be realised, conceptually this can only be achieved through reducing the number of sheep and beef cattle. At this point the farming and policy worlds again merge, but in ways that are threatening to the livelihoods of farmers.

# 11. CONCLUSIONS

This paper extends the concept of co-production as an analytical framework for understanding the evaluation of a 'good' animal to encompass methane. As noted earlier, the sheep and beef farming sector tends not to be strongly influenced by technology (although exceptional farms exist).The construction of an animal is predicated on a range of other factors, in particular the environment and consumer market orientation. However, across a diverse range of farming approaches, it became very apparent that farmers have difficulty in accepting the role of sheep and beef cattle in climate change. Methane emissions were viewed as normal and natural.

The policy imperative to reduce greenhouse gas emissions has resulted in the identification of a number of technical approaches to reducing methane, requiring action by farmers. These technological approaches presume that the animal can be separated from its methane emissions. However, from farmers' perspectives, methane emissions are an integral part of the animal and, therefore, difficult to conceptualise as amenable to technological manipulation.

Unlike most pigs and chickens which tend to be kept in uniform environmental conditions, sheep and beef cattle farming is characterised by heavy reliance on grazing grass, often in challenging climatic and topographical situations. The environment, therefore, becomes an important part of the construction of sheep and beef cattle. Animals are literally selected for survival by their environments. This local and specific conceptualisation of animals contrasts with the presumption that technologies can be applied across all environments. Technologies that reduce methane emissions by increasing productivity are synergistic with farming systems where efficiency of production is emphasised only as long as other values and aspirations are not compromised. Methane reduction has tended to become associated with paradigms of industrialised agriculture emphasising the utilitarian functions of animals and this is anathema to farmers aiming at more holistic approaches.

The range of different markets to which the meat animal is sold also has a strong impact on the definition of the 'good' animal; with emphasis on naturalness and links to the local for sales direct to consumers, and emphasis on shape and fatness for sales to slaughterhouses. Consumer responsibilities have largely been limited to exhortations to reduce consumption of red meat. The moral responsibility has thus largely been shifted from consumers to producers, with little evidence of demand from consumers to apply technology to animals to reduce methane emissions.

The sheep and beef cattle context involves complex assemblages of animals, nature, markets and technologies to providing a portfolio of different social and private 'goods' depending on the farmers own values and circumstances. This paper identifies exemplars of how contextual factors dramatically impact on animal co-productions highlighting a serious mismatch between policy/scientific constructions and farmer constructions. I suggest that to include methane in the category 'manipulable by technology' requires more sophisticated co-construction of the animal and technology. While application of technology is one possible policy approach to reducing methane, alternative methods, such as reducing demand for meat, or balancing stocking rates with carbon absorption capacity, need not be neglected.

## 12. ENDNOTES

[a]Numbering is retained from automatic numbering on the recording device in order to maintain traceability of evidence. Hence the numbers allocated do not start at 0 and therefore exceed 42.

## AUTHOR'S INFORMATION

AB is a Research Fellow at the University of Edinburgh holding post-graduate degrees in both innovation studies and animal genetics. As well as lecturing, she is currently employed as an Agri-Food Knowledge-Exchange Fellow with the Natural Environment Research Council.

## ACKNOWLEDGEMENTS

The research on which this paper is based was supported by grant RES-000-22-3737 from the Economic and Social Research Council and forms part of the research programme at the ESRC Innogen Centre at the University of Edinburgh. I am grateful for all the interviewees who kindly took part in this research and for members of my advisory group. Many thanks also to David Wield, Sue Oreszczyn, Isabel Fletcher, Donald Bruce, Elisabeth Barlow and two anonymous reviewers for comments on an earlier draft and to Eileen Mothersole for transcribing interviews.


## COMPETING INTERESTS

The support of the Centre for Society and the Life Sciences, Radboud University, the Netherlands and Economic and Social Research Council Genomics Network UK for publication costs is gratefully acknowledged.

## REFERENCES


1 Barnes AP, Toma L: A typology of dairy farmer perceptions towards climate change. Climate Change 2012, 112: 507–522.

2 Barr S, Shaw G, Coles T: Sustainable lifestyles: sites, practices and policy. Environ Plann A 2011, 43: 3011–3029.

3 Blackstock KL, Ingram J, Burton R, Brown KM, Slee B: Understanding and influencing behaviour change by farmers to improve water quality. Sci Total Environ 2010, 408: 5631–5638.

4 Burton RJF, Kuvzera C, Schwarz G: Exploring Farmer's cultural resistance to voluntary agri-environmental schemes. Sociol Rural2008,48(1):16–37.

5 Calvert S: Certified angus, certified patriot: breeding, bodies and pedigree practices. Sci Culture 2013,22(3):291–313.

6 Committee on Climate Change: The Fourth Carbon Budget. Reducing emissions through the 2020s. London; 2010. . Accessed 16 June 2012 http://www.theccc.org.uk/publication/the-fourth-carbon-budget-reducing-emissions-through-the-2020s-2/.

7 Convery IT, Bailey C, Mort M, Baxter J: Death in the wrong place? emotional geographies of the UK 2001 foot and mouth disease epidemic. J Rural Stud 2004,21(1):99–109.

8 Doubleday R: Organising accountability: co-productions of technoscientific and social worlds in a nanoscience laboratory. Area2007,39(2):166–175.

9 Douglas M: Purity and Danger: An analysis of the Concepts of Pollution and Taboo. London: ARK paperbacks; 1966.

10 Eden S, Bear C, Walker G: Mucky carrots and other proxies: problematising the knowledge-fix for sustainable and ethical consumption. Geoforum 2008, 39: 1044–1057.

11 Evans N, Gaskell P, Winter M: Re-assessing agrarian policy and practice in local environmental management: the case of beef cattle. Land Use Policy 2003, 20: 231–242.

12 Foresight. The Future of Food and Farming: Final Project Report. London: The Government Office for Science; 2011.

13 Garnett T: Where are the best opportunities for reducing greenhouse gas emissions in the food system (including the food chain)? Food Policy 2011, 36: S23-S32.

14 Garnett T, Godfray C: Sustainable intensification in agriculture. Navigating a course through competing food system priorities. UK: Food Climate Research Network and the Oxford Martin Programme on the Future of Food, University of Oxford; 2012. . Accessed 10 Feb 2013http://www.fcrn.org.uk/fcrn/publications/sustainable-intensification-agriculture.

15 Genesis Faraday Partnership: A study of the scope for the application of research in animal genomics and breeding to reduce nitrogen and methane emissions from livestock based food chains. Defra research project AC204; 2008. final report. http://randd.defra.gov.uk/ Default.

aspx?Menu=Menu&Module=More&Location=None&Completed=0&ProjectID=14662.

16 Gibbs D, Holloway L, Gilna B, Morris C: Genetic techniques for livestock breeding: restructuring institutional relationships in agriculture. Geoforum 2009, 40: 1041–1049.

17 Gilg A, Barr S, Ford N: Green consumption or sustainable lifestyles? Identifying the sustainable consumer. Futures 2005, 37: 481–504.

18 Gill M, Smith P, Wilkinson JM: Mitigating climate change: the role of domestic livestock. Animal 2010,4(3):323–333.

19 Goodman D: Agro-food studies in the 'Age of Ecology': nature, corporeality, Bio-politics. Sociol Rural 1999,39(1):18–38.

20 Grasseni C: Managing cows: an ethnography of breeding practices and uses of reproductive technology in contemporary dairy farming in Lombardy (Italy). Stud Hist Philos Biol Biomed Sci 2007, 38: 488–510.

21 Gray J: Family farms in the Scottish borders: a practical definition by hill sheep farmers. J Rural Stud 1998,14(3):341–356.

22 Green K, Foster C: Give peas a chance: transformations in food consumption and production systems. Technol Forecasting Soc Change 2005, 72: 663–679.

23 Haden VR, Niles MT, Lubell M, Perlman J, Jackson LE: Global and Local Concerns: What Attitudes and Beliefs Motivate Farmers to Mitigate and Adapt to Climate Change?. 2012.

24 Haggerty J, Campbell H, Morris C: Keeping the stress off the sheep? Agricultural intensification neoliberalism, and 'good' farming in New Zealand. Geoforum 2009, 40: 767–777.

25 Holloway L: Understanding climate change and farming: scientific and farmers' constructions of 'global warming' in relation to agriculture. Environ Plann A 1999, 31: 2017–2032.

26 Holloway L: Aesthetics, genetics, and evaluating animal bodies: locating and displacing cattle on show and in figures. Environ Plann: Soc Space 2004, 23: 883–902.

27 Holloway L: Subjecting cows to robots: farming technologies and the making of animal subjects. Environ Plann D: Soc Space2007, 25: 1041–1060.

28 Holloway L, Bear C: Commentary. Environ Plann A 2011, 43: 1487–1491.

29 Holloway LE, Ilbery BW: Famers' attitudes towards environmental change, particularly global warming, and the adjustment of crop mix and farm management. Appl Geogr 1996,16(2):159–171.

30 Holloway L, Morris C: Boosted bodies: genetic techniques, domestic livestock bodies and complex representations of life.Geoforum 2008, 39: 1709–1720.

31 Holloway L, Morris C, Gilna B, Gibbs D: Biopower, genetics and livestock breeding: (re)constituting animal populations and heterogeneous biosocial collectivities. Trans Ins British Geo NS 2009, 34: 394–407.

32 Holloway L, Morris C, Gilna B, Gibbs D: Choosing and rejecting cattle and sheep: changing discourses of practices of (de)selection in pedigree livestock breeding. Agric Hum Values 2011,28(4):533–547.

33 Holstein JA, Gubrium JF: Active Interviewing. In Qualitative Research Methods. Edited by: Weinberg D. Oxford and Malden: Blackwell Publishers; 2002:112–126.

34 Industry Delivery Partners: Meeting the Challenge: Agriculture Industry GHG Action Plan Phase 1: 2010–2012 4th April. 2011. . Accessed 16 June 2012 http://www.ahdb.org. uk/projects/GreenhouseGasActionPlan.aspx . Accessed 16 June 2012

35 Jasanoff S: The idiom of co-production. In States of Knowledge. The co-production of science and social order. Edited by: Jasanoff S. London: Routledge; 2006:1–12.

36 Jones AK, Jones DL, Edward-Jones G, Cross P: Informing decision making in agricultural greenhouse gas mitigation policy: a best-worst scaling survey of expert and farmer opinion in the sheep industry. Environ Sci Pol 2013, 29: 46–56.

37 Kahneman D, Tversky A: Judgement under Uncertainty. Cambridge: Cambridge University Press; 1982.

38 Lowe P, Cark J, Seymour S, Wars N: Moralizing the Environment: Countryside Change, Farming and Pollution. London: UCL Press; 1997.

39 Macnaghten P: Animals in their nature: a case study on public attitudes to animals, genetic modification and 'Nature'. Sociology2004, 38: 533–551.

40 Moran D, Macleod M, Wall E, Ivery V, McVittie A, Barnes A, Rees R, Topp CFE, Moxey A: Marginal abatement cost curves for UK agricultural greenhouse Gas emissions. J Agric Econ 2011,62(1):63–118.

41 Murdoch J, Marsden T, Banks J: Quality, nature and embeddedness: some theoretical considerations in the context of the food sector. Econ Geogr 2000,76(2):107–125.

42 Reardon J: The human genome diversity project: a case study in coproduction. Soc Stud Sci 2001, 31: 357–388.

43 Riley M: 'Letting them go' – agricultural retirement and human-livestock relations. Geoforum 2011, 42: 16–27.

44 Ritchie J, Lewis J, Elam G: Designing and Selecting Samples. In Qualitative Research Practice. A Guide for Social Science Students and Researchers. Edited by: Ritchie J, Lewis J. London: Sage Publications; 2003:77–108.

45 Saltzman K, Head L, Stenseke M: Do cows belong in nature? The cultural basis of agriculture in Sweden and Australia. J Rural Stud 2011, 27: 54–62.

46 Schot J, Rip A: The past and future of constructive technology assessment. Technol Forecast Soc Chang 1996, 54: 251–268.

47 Spence A, Pidgeon N: Framing and communicating climate change: the effects of distance and outcome frame manipulations.Glob Environ Chang 2010, 20: 656–667.

48 Spence A, Poortinga W, Pidgeon N, Lorenzoni I: Public perceptions of energy choices: the influence of beliefs about climate change and the environment. Energy Environ 2010,21(5):385–407.

49 Stassart PM, Whatmore SJ: Metabolising risk: food scares and the un/re-making of Belgian beef. Environ Plann A 2003, 35: 449–462.

50 Steinfeld H, Gerber P, Wassenaar T, Castel V, Rosales M, de Haan C: Livestock's Long Shadow: environmental issues and options. Rome: Food and Agriculture Organization of the United Nations; 2006. ftp://ftp.fao.org/docrep/fao/010/a0701e/a0701e00.pdf. Accessed 15 June 2012.

51 Ward N, Lowe P, Seymour S, Clark J: Rural restructuring and the regulation of farm pollution. Environ Plann A 1995, 27: 1193–1211.

52 Ward N, Clark J, Lowe P, Seymour S: Keeping matter in its place: pollution regulation and the reconfiguring of farmers and farming. Environ Plan 1998, 30: 1165–1178.

53 Wilkie R: Sentient commodities and productive paradoxes: the ambiguous nature of human-livestock relations in Northeast Scotland. J Rural Stud 2005, 21: 213–230.

54 Yarwood R, Evans N: Livestock, locality and landscape: EU regulations and the new geography of Welsh farm animals. Appl Geogr 2003, 23: 137–157.

55 Yarwood R, Evans N: A Lleyn sweep for local sheep? Breed societies and the geographies of Welsh livestock. Environ Plann A2006, 38: 1307–1326.

# CHAPTER 4

# Genetic diversity and relationship of cattle populations of East India: distinguishing lesser known cattle populations and established breeds based on STR markers

**Rekha Sharma, Avishek Maitra, Pramod Kumar Singh and Madhu Sudan Tantia**

National Bureau of Animal Genetic Resources, Karnal 132 001, Haryana, India

## ABSTRACT

India has 34 recognized breeds of cattle in addition to many more not characterized and accredited so far. It is imperative to characterize all the cattle germplasm of the country so as to have better breeding and conservation options. Thus, present study was planned for assessing genetic diversity and relationship between three local cattle populations

(Gangatiri, Shahabadi and Purnea) and two established cattle breeds (Bachaur and Siri) of eastern India by using 21 FAO and ISAG recommended microsatellite markers. A total of 243 unrelated DNA samples of five cattle populations were collected from respective habitats. A total of 304 microsatellite alleles were identified with number of alleles at one locus ranging from 5 to 29. The average observed heterozygosity lie within the narrow range of $0.681 \pm 0.04$ in Purnea to $0.721 \pm 0.03$ in Siri. Mean estimates of observed and expected heterozygosity over all loci and breeds were $0.704 \pm 0.02$ and $0.720 \pm 0.01$, respectively. In the overall population, the homozygote excess ($F_{IT}$) of $0.073 \pm 0.02$, was partly due to the homozygote excess within breeds ($F_{IS} = 0.026 \pm 0.02$) and to a larger extent due to genetic differentiation among breeds ($F_{ST} = 0.048 \pm 0.01$). The genetic distance, STRUCTURE and Principal Component Analyses concluded that the Siri cattle are most distinct among the investigated cattle populations. Furthermore the analysis of genetic structure indicated that the most probable number of clusters is four. All analysis showed that a significant amount of genetic variation is maintained in local cattle populations of which Shahabadi and Purnea are distinct from the recognized breeds of the area and needs recognition as breeds.

## KEYWORDS

Conservation Diversity Genetic relationship Indian Cattle Microsatellite markers Population structure

## 1. INTRODUCTION

A large and divergent range of agro-ecological zones in India has helped to develop number of cattle populations. Cattle are the largest livestock species in India and constitute 37.5% of its total livestock population (BAHS-Basic Animal Husbandry Statistics 2012). However, it is surprising to note that only 11.6% of total cattle (199.08 million) belong to pure indigenous breeds, whereas, 69.7% were classified as non-descript indigenous animals (BAHS-Basic Animal Husbandry

Statistics 2012). Non-descript population includes the cross-bred populations, populations which are mixture of different breeds or the populations which have not yet been studied or described. Non-descript population is greatly contributing to the total Gross Domestic Product from livestock sector. In addition, indigenous animal genetic resources is known for heat and draught tolerance, disease resistance and subsistence on poor feed, opening scope for allele mining for these traits. Thus the emphasis should be to describe, characterize and document lesser known populations in the country so that the proportion of non-descript population is considerably transformed in to defined breeds. This will be first step towards planning the organized breeding program for their genetic improvement, conservation strategies and sustainable utilization.

Gangatiri, Shahabadi and Purnea are three non-descript cattle populations of eastern India. The registered cattle breeds in and surrounding area include Bachaur and Siri. Bachaur, Shahabadi and Purnea cattle are from Bihar state whereas, Gangatiri belongs to area of Uttar Pradesh state bordering Bihar and habitat of Siri borders Bihar state on eastern side (Figure 1). Typical animals of these breeds and populations are available in their native tracts in optimum number (>4000) to constitute viable breeding populations (FAO 2007). Since last two decades these populations are showing declining trend (Singh 2009) thus need immediate attention, failing which our ancestral efforts of artificial selection, operative since time immemorial, may go in vein and there may be a non-repairable loss from the national and world gene pool. Employment of microsatellite markers is one of the most powerful means for studying the genetic diversity, calculation of genetic distances, detection of bottlenecks and admixture because of their high degree of polymorphism, random distribution across the genome and neutrality with respect to selection (Dcdgson *et al.*1997). Considering the importance of cattle in Indian agriculture, efforts have been made to evaluate the genetic diversity and relationship in Indian cattle using microsatellite markers. These included native cattle breeds adapted to the north-western arid and semi-arid region of India (Sodhi *et al.*2008; 2011), north Indian cattle breeds (Sharma *et al.*2009), South Indian breeds (Metta *et al.*2004) and cattle of Orissa state and hill cattle of Kumaun (Sharma *et al.*2012a). However, the genetic relationship between native cattle breeds of eastern India is unknown.

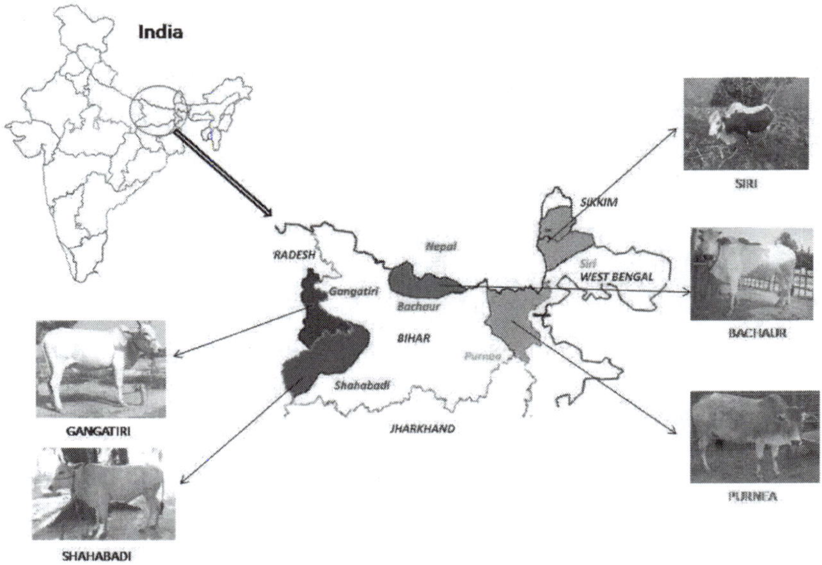


**Figure 1** Distribution of cattle populations.

The present study aimed to genetically characterize and distinguish lesser known cattle populations (Shahabadi, Purnea and Gangatiri) and recognized cattle breeds of eastern India (Bachaur and Siri). Twenty one microsatellites (Short Tandem Repeats, STR) were amplified in five multiplex PCR. We intend to evaluate not only the current diversity but also to know their relationship for the conservation of the genetic diversity in the context of biodiversity management programs.

# 2. MATERIALS AND METHODS

## 2.1 Biological material

Shahabadi, Purnea and Gangatiri are non-descript cattle from Eastern India (Figure 1). Gangatiri and Shahabadi are white or grey colored animals. Gangatiri animals are reared for milk production (4–6 liters per day) as well as for agricultural operations (Singh *et al.*2007). Shahabadi cattle are mainly distributed in Buxar, Bhojpur, Kaimoor and Rohtas (Sasaram) districts of Bihar. The milk production varies from 2 to 6 liters in a day (Sharma *et al.*2012b). Purnea animals are of small size and compact body. The coat colour of Purnea cattle varies from light to deep red with few greyish white animals (Sharma *et al.*2013). The animals are primarily employed for agricultural operations, for carrying load and transportation. The habitat of Purnea cattle encompasses Purnia commissionery of Bihar state (Purnia, Araria, Katihar and Kishanganj districts).

Bachaur and Siri are recognized cattle breeds of Eastern India. Bachaur, a famous draught purpose cattle breed is concentrated in Bachaur Pargana (Madhubani, Sitamarhi and Darbhanga districts) of Bihar state. The habitat of Bachaur has shrunken with time to northern part of Bihar bordering Nepal. The animals are white to grey in colour. The cows are poor milker, producing an average of 2.2 kg of milk per day and are managed under extensive management (Sharma *et al.*2007). Animals of Siri breed are small in size and are distributed in the hilly tracts of West Bengal (adjoining state of Bihar) and Sikkim states of India. The color most frequently seen is black and white or extensive solid black. The Siri has a hump that is thoracic, muscular-fatty and slightly forward in position as compared to other Zebu breeds of India. The animal carries a thick coat all the year around which protects them from heavy rains and severe cold (Sharma *et al.*2008).

## 2.2 Blood sampling and microsatellite analysis

Samples of the populations included in this study represented animals of the original autochthonous phenotype. Blood samples from 243 individuals were collected (Bachaur-50, Gangatiri-50, Shahabadi-48, Purnea-47 and Siri-48) from different villages of habitat (Figure 1) while avoiding closely related individuals on the basis of detailed interview with owners. Blood samples collected in 10 ml vacuitainer tubes containing EDTA as anticoagulant were stored at −20°C until DNA extraction. DNA isolation was carried out using modified Phenol-chloroform method (Sambrook *et al.*1989). Isolated DNA samples were amplified by PCR in correspondence with the selected panel of 21 loci. The loci were chosen, according to ISAG/FAO recommendation aiming to analyze high polymorphic markers spread all over the genome and ability to co-amplify in PCR reactions (FAO [Food and Agricultural

Organization of the United Nations] 2011). The fluorochrome labeled (FAM, NED, PET& VIC) primers were synthesized by Applied Biosystems (Table 1). For amplification, 50-100 ng of genomic DNA was added to a reaction mixture containing 50 pMol of primer- forward and reverse, 200 μM of every dNTPs, 1.5 mM of $MgCl_2$ and 0.5U of *taq* polymerase in a final volume of 25 μl. All the microsatellites were amplified by a MJ thermal cycler at the following conditions: initial denaturation step of 1 min at 95°C, 30 cycles of 1 min at 95°C, 1 min at T°C (optimum annealing temperature of each primer) and 1 min at 72°C and a final extension of 5 min at 72°C. Amplified fragments were separated by capillary electrophoresis using an ABI PRISM 310 automatic sequencer (Applied Biosystems, Foster City, CA, USA). Fluorescently labeled fragments were detected and sized using GeneMapper (version 3.7, Applied Biosystems).

**Table 1** Characteristics of 21 microsatellite loci used in present study

| Primers | Primer sequences (5'-3') | Forward label | Set | Annealing temp | Product size (bp) | Total number of alleles per locus |
|---|---|---|---|---|---|---|
| BM1824 | gagcaaggtgttttccaatc<br>cattctccaactgcttccttg | VIC | 4 | 58°C | 178-194 | 7 |
| CSSM08 | cttggtgttactagccctggg<br>gatatatttgccagagattctgca | VIC | 3 | 55°C | 182-200 | 7 |
| CSSM33 | cactgtgaatgcatgtgtgtgagc<br>cccatgataagagtgcagatgact | NED | 5 | 58°C | 148-186 | 19 |
| CSSM66 | acacaaatcctttctgccagctga<br>aatttaatgcactgaggagcttgg | FAM | 4 | 60°C | 167-207 | 17 |
| ETH10 | gttcaggactggccctgctaaca<br>cctccagcccactttctcttctc | NED | 1 | 58°C | 185-221 | 14 |
| ETH225 | gaacctgcctctcctgcattgg<br>actctgcctgtggccaagtagg | VIC | 4 | 64°C | 134-156 | 11 |
| ETH3 | gatcaccttgccactatttcct<br>acatgacagccagctgctact | NED | 4 | 57°C | 92-122 | 14 |
| HEL09 | cccattcagtcttcagaggt<br>cacatccatgttctcaccac | FAM | 5 | 59°C | 140-168 | 13 |
| HEL5 | gcaggatcacttgttaggga<br>agacgttagtgtacattaac | VIC | 3 | 55°C | 137-187 | 20 |

| | | | | | | |
|---|---|---|---|---|---|---|
| ILSTS11 | gcttgctacatggaaagtgc<br>ctaaaatgcagagccctacc | NED | 1 | 58°C | 261-269 | 5 |
| ILSTS34 | aagggtctaagtccactggc<br>gacctggtttagcagagagc | VIC | 5 | 59°C | 138-208 | 29 |
| ILSTS33 | tattagagtggctcagtgcc<br>atgcagacagttttagaggg | PET | 3 | 55°C | 137-163 | 10 |
| INRA05 | caatctgcatgaagtataaatat<br>cttcaggcataccctacacc | FAM | 2 | 54°C | 130-144 | 8 |
| INRA35 | atcctttgcagcctccacattg<br>ttgtgctttatgacactatccg | FAM | 3 | 54°C | 80-142 | 24 |
| INRA63 | atttgcacaagctaaatctaacc<br>aaaccacagaaatgcttggaag | PET | 2 | 54°C | 164-188 | 11 |
| MM12 | caagacaggtgtttcaatct<br>atcgactctggggatgatgt | PET | 4 | 52°C | 88-132 | 20 |
| MM8 | cccaaggacagaaaagact<br>ctcaagataagaccacacc | NED | 2 | 55°C | 114-140 | 10 |
| TGLA122 | ccctcctccaggtaaatcagc<br>aatcacatggcaaataagtacatac | VIC | 1 | 58°C | 135-179 | 18 |
| TGLA227 | cgaattccaaatctgttaatttgct<br>acagacagaaactcaatgaaagca | PET | 2 | 55°C | 97-119 | 16 |
| TGLA53 | gctttcagaaatagtttgcattca<br>atcttcacatgatattacagcaga | FAM | 1 | 58°C | 142-184 | 20 |

## 2.3 Statistical analysis

GENALEX 6.2 software (Peakall and Smouse, 2008) was used to estimate basic population genetic descriptive statistics for each marker and population: gene frequency, observed number of alleles ($N_o$), number of private alleles, effective number of alleles ($N_e$), observed ($H_o$) and expected heterozygosity ($H_e$) and Hardy-Weinberg equilibrium (HWE). Wright's statistics $F_{IS}$ (f), $F_{ST}$ (θ) and $F_{IT}$ (F) and Nei's (Nei 1987) standard genetic distances among populations were calculated as implemented in GENALEX software. Pair wise matrix of the genetic distances was then used to obtain a Neighbor-joining tree which was visualized using the software TreeView (Page 1996). Bootstraps of 1000 replicates were performed in order to test the robustness of tree topology using the Phylip software (Felsenstein 1993). Multivariate analysis of microsatellite allele frequencies (Principal Component analysis, PCA) was applied to reveal the underlying evolutionary history and admixture among populations. An alternative model-based Bayesian clustering analysis was used to infer how many clusters or subpopulations (K) were most appropriate for interpreting the data without prior information on the number of locations at which the individuals were sampled as implemented in STRUCTURE v2.2 (Pritchard *et al.*2000). Analysis was performed with a burn-in length of 50,000 followed by 30,000 MCMC (Marcov Chain Monte Carlo) iterations for each K = 1 to 7 with five replicate runs for each K using independent allele frequencies and no admixture. Optimal K value was selected after analyzing the result files with STRUCTURE Harvester (Earl and vonHoldt2012).

# 3. RESULTS AND DISCUSSION

In the present study genetic status and diversity of lesser known indigenous cattle populations of eastern India and their relationship with established breeds of the same region was established using microsatellite markers. All microsatellite markers used in this study were successfully amplified in five multiplex sets designed with consideration for annealing temperature, product size and specific dye label in all the populations (Table 1). The genotype data generated in present study showed that significant amount of genetic variation is maintained in local cattle populations. All the markers were found to be polymorphic in each of the five populations analyzed. Considering all the populations, majority of the markers were in HWE. Number of loci deviating from HWE ($P < 0.05$) were one for Gangatiri (ETH3) and Shahabadi (MM8), two for Bachaur (TGLA122, MM12), three for Siri (TGLA227, ETH10, CSSM66) and none for Purnea cattle populations. The level of variations depicted by number of alleles at each locus serves as a measure of genetic variability having direct effect on differentiation of breeds within a species (Buchanan et al. 1994). All the 21 microsatellite loci showed ample polymorphism for evaluating within breed genetic variability and exploring genetic differences between breeds. A total of 304 alleles were detected with ILSTS34 presenting the highest number of alleles per locus (29) while ILSTS11 presented the lowest (5) number of alleles (Table 1). ILSTS34 presented the highest number of alleles per locus (19) in Bachaur cattle while BM1824, ETH10 and ILSTS11 presented the lowest (4) in Purnea cattle. The effective number of alleles per locus in a population varied from 1.492 (TGLA227) to 10.301 (INRA35) and was proportional to the value of expected heterozygosity found in these loci (TGLA227, 0.330 and INRA35, 0.903) (Table 2). Lower values of expected number of alleles as compared to observed number of alleles in all the populations suggested that there were many low frequency alleles in the populations. The mean observed number of alleles across all the loci was 9.486 ± 0.327 and was higher than other indigenous cattle breeds (Metta *et al.*2004; Mukesh *et al.*2004; Pandey *et al.*2006a and2006b). Lower allelic diversity than studied populations have also been reported in exotic cattle- Burlina-6.7 (Dalvit *et al.*2008), Brown Swiss-5.4 (Schmid *et al.*1999) and Creole cattle-7.2 (Egito *et al.*2007). Previously also the allelic diversity in the Indian livestock breeds has been observed to be higher than that reported for the European counterpart (Joshi *et al.*2012) that has been attributed to lack of artificial selection pressure. Allelic diversity of similar magnitude has also been reported in Tharparkar, Rathi and Orissa cattle populations of India (Sodhi *et al.*2008; Sharma *et al.*2012a). Measures of genetic diversity based on allelic richness are considered important in conservation genetics as marker-assisted methods for maximizing number of alleles conserved have been shown to be effective (Bataillon *et al.*1996). It is also relevant in long-term perspective, as selection limits are determined by the initial allelic composition rather than by heterozygosity (Petit et al.1998).

Estimates of observed heterozygosity including all loci and population (0.704 ± 0.016) confirmed the remarkable level of diversity in the studied populations. Among populations, observed heterozygosity ranged from 0.681 ± 0.04 to 0.721 ± 0.028 with the lowest value found in Purnea cattle and the highest in Siri cattle. Most of the indigenous breeds including Kherigarh-0.57 (Pandey *et al.*2006a), Kenkatha-0.54 (Pandey *et al.*2006b), Sahiwal-0.43 (Mukesh *et al.*2004) and Deoni-0.59 (Metta *et al.*2004) showed lower estimates of observed heterozygosity than local cattle populations and the breeds investigated in present study. Overall heterozygosity estimates were comparable with Tharparkar cattle (0.64, Sodhi *et al.*2008), Orissa cattle populations (0.62 to 0.66, Sharma *et al.*2012a) of India, Chinese cattle (0.62, Sun *et al.*2008) and Creole cattle (0.61, Egito *et al.*2007). Purnea cattle represented the lowest observed and effective number of alleles as well as observed and expected heterozygosity. On the other hand Siri cattle presented the highest value of the above said parameters amongst all genetic groups (Table 2). Genetic variation is necessary to allow organisms to adapt to ever changing environments with some of this variation stemming from introduction of new alleles by the random and natural process of mutation. Higher genetic variation in Siri cattle must have contributed to its adaptability and this is reflected in wide-spread distribution of Siri cattle not only in India but also in neighboring country, Bhutan.

**Table 2** Genetic variability parameters of five cattle populations

| Locus | Bachaur | | | | Gangatiri | | | | Shahabadi | | | | Purnea | | | | Siri | | | |
|---|---|---|---|---|---|---|---|---|---|---|---|---|---|---|---|---|---|---|---|---|
| | No | Ne | Ho | He | No | Ne | Ho | He | No | Ne | Ho | He | No | Ne | Ho | He | No | Ne | Ho | He |
| **BM1824** | 6 | 2.678 | 0.580 | 0.627 | 5 | 2.400 | 0.660 | 0.583 | 6 | 2.471 | 0.532 | 0.595 | 4 | 2.297 | 0.404 | 0.565 | 6 | 3.683 | 0.667 | 0.729 |
| **CSSM08** | 7 | 2.119 | 0.510 | 0.528 | 6 | 2.533 | 0.667 | 0.605 | 6 | 2.496 | 0.545 | 0.599 | 7 | 2.409 | 0.581 | 0.585 | 6 | 2.950 | 0.600 | 0.661 |
| **CSSM33** | 14 | 6.767 | 0.875 | 0.852 | 15 | 6.840 | 0.860 | 0.854 | 16 | 7.810 | 0.896 | 0.872 | 13 | 7.007 | 0.826 | 0.857 | 12 | 5.914 | 0.830 | 0.831 |
| **CSSM66** | 9 | 3.740 | 0.680 | 0.733 | 11 | 4.803 | 0.860 | 0.792 | 13 | 5.086 | 0.854 | 0.803 | 14 | 5.084 | 0.745 | 0.803 | 8 | 3.411 | 0.688 | 0.707 |
| **ETH10** | 9 | 4.655 | 0.880 | 0.785 | 11 | 5.025 | 0.900 | 0.801 | 6 | 4.064 | 0.814 | 0.754 | 4 | 2.673 | 0.891 | 0.626 | 14 | 7.266 | 0.894 | 0.862 |
| **ETH225** | 9 | 2.231 | 0.560 | 0.552 | 8 | 2.019 | 0.500 | 0.505 | 10 | 2.051 | 0.583 | 0.512 | 10 | 6.258 | 0.851 | 0.840 | 9 | 4.148 | 0.708 | 0.759 |
| **ETH3** | 8 | 3.541 | 0.980 | 0.718 | 9 | 3.671 | 0.860 | 0.728 | 14 | 6.508 | 0.958 | 0.846 | 7 | 3.899 | 0.979 | 0.744 | 8 | 3.796 | 0.979 | 0.737 |
| **HEL09** | 10 | 7.529 | 0.979 | 0.867 | 10 | 6.527 | 0.900 | 0.847 | 10 | 7.396 | 0.896 | 0.865 | 10 | 8.000 | 0.833 | 0.875 | 12 | 8.243 | 0.766 | 0.879 |
| **HEL5** | 11 | 4.361 | 0.714 | 0.771 | 10 | 5.219 | 0.857 | 0.808 | 13 | 6.585 | 1.000 | 0.848 | 10 | 3.901 | 0.844 | 0.744 | 10 | 6.853 | 0.756 | 0.854 |
| **ILSTS06** | 7 | 3.058 | 0.500 | 0.673 | 8 | 4.062 | 0.640 | 0.754 | 9 | 3.298 | 0.545 | 0.697 | 8 | 4.691 | 0.619 | 0.787 | 9 | 4.485 | 0.489 | 0.777 |
| **ILSTS11** | 5 | 2.288 | 0.469 | 0.563 | 5 | 2.600 | 0.500 | 0.615 | 5 | 3.150 | 0.558 | 0.683 | 4 | 1.928 | 0.422 | 0.481 | 5 | 4.212 | 0.638 | 0.763 |
| **ILSTS34** | 19 | 7.444 | 0.875 | 0.866 | 16 | 6.046 | 0.760 | 0.835 | 15 | 4.571 | 0.750 | 0.781 | 10 | 4.054 | 0.652 | 0.753 | 14 | 6.914 | 0.851 | 0.855 |
| **ILSTS33** | 8 | 3.460 | 0.714 | 0.711 | 7 | 2.350 | 0.548 | 0.575 | 9 | 4.150 | 0.727 | 0.759 | 6 | 2.864 | 0.531 | 0.651 | 8 | 4.295 | 0.733 | 0.767 |
| **INRA05** | 7 | 4.367 | 0.640 | 0.771 | 8 | 4.394 | 0.820 | 0.772 | 7 | 4.509 | 0.708 | 0.778 | 6 | 4.251 | 0.581 | 0.765 | 7 | 3.938 | 0.729 | 0.746 |
| **INRA35** | 12 | 5.723 | 0.755 | 0.825 | 8 | 4.570 | 0.667 | 0.781 | 17 | 10.301 | 0.860 | 0.903 | 6 | 3.081 | 0.621 | 0.675 | 8 | 4.054 | 0.844 | 0.753 |
| **INRA63** | 6 | 2.501 | 0.520 | 0.600 | 5 | 2.627 | 0.580 | 0.619 | 6 | 2.781 | 0.563 | 0.640 | 7 | 3.446 | 0.837 | 0.710 | 7 | 2.366 | 0.521 | 0.577 |
| **MM12** | 11 | 5.149 | 0.760 | 0.806 | 10 | 4.146 | 0.720 | 0.759 | 13 | 4.295 | 0.729 | 0.767 | 15 | 4.909 | 0.787 | 0.796 | 10 | 2.841 | 0.604 | 0.648 |
| **MM8** | 8 | 2.737 | 0.560 | 0.635 | 6 | 2.667 | 0.680 | 0.625 | 8 | 3.072 | 0.625 | 0.674 | 8 | 2.703 | 0.605 | 0.630 | 6 | 2.525 | 0.583 | 0.604 |
| **TGLA122** | 14 | 8.666 | 0.900 | 0.885 | 12 | 9.728 | 0.920 | 0.897 | 14 | 9.732 | 0.791 | 0.897 | 13 | 7.627 | 0.844 | 0.869 | 12 | 8.925 | 0.915 | 0.888 |
| **TGLA227** | 6 | 1.492 | 0.380 | 0.330 | 8 | 1.571 | 0.360 | 0.363 | 12 | 1.831 | 0.417 | 0.454 | 10 | 1.871 | 0.302 | 0.466 | 11 | 2.869 | 0.667 | 0.651 |
| **TGLA53** | 13 | 3.408 | 0.740 | 0.707 | 15 | 2.648 | 0.640 | 0.622 | 14 | 3.485 | 0.628 | 0.713 | 15 | 2.557 | 0.545 | 0.609 | 12 | 3.532 | 0.681 | 0.717 |
| **Mean** | **9.476** | **4.186** | **0.694** | **0.705** | **9.190** | **4.117** | **0.709** | **0.702** | **10.619** | **4.745** | **0.713** | **0.735** | **8.905** | **4.072** | **0.681** | **0.706** | **9.238** | **4.629** | **0.721** | **0.751** |
| **S. D.** | **0.752** | **0.440** | **0.038** | **0.030** | **0.716** | **0.436** | **0.034** | **0.030** | **0.824** | **0.532** | **0.035** | **0.027** | **0.771** | **0.402** | **0.040** | **0.027** | **0.589** | **0.421** | **0.028** | **0.020** |

Observed heterozygosity was lower than expected heterozygosity in Bachaur, Shahabadi, Purnea and Siri cattle populations, showing departure from HWE and analysis of $F_{IS}$ evidenced some heterozygote deficiency too (Table 3). This disequilibrium was caused by heterozygote deficiency in each population which was highest in Purnea ($0.042 \pm 0.041$) and lowest in Bachaur ($0.017 \pm 0.030$). On the contrary, Gangatiri cattle presented slight heterozygote excess in the population ($-0.010 \pm 0.022$) which was expressed in heterozygosity pattern too ($H_o = 0.709 \pm 0.034$, $H_e = 0.702 \pm 0.030$) (Table 2). Existence of this population in small geographical region with free grazing of Gangatiri and non-descript animals in a herd could be the likely sources for the sufficient heterozygotes. Positive $F_{IS}$ estimate for Bachaur, Shahabadi, Purnea and Siri indicated either the presence of inbreeding and /or Wahlund effect (presence of population substructure within breed). Since blood samples were collected from different villages, presence of a hidden substructure cannot be ruled out. In fact, animals of the same population but belonging to different villages could derive from genetically different founders. A significant homozygote excess was also observed in other studies on indigenous cattle breeds such as Sahiwal $F_{IS} = 0.32$ (Mukesh *et al.*2004). Most likely inbreeding in these populations is arising from unplanned and unsystematic breeding owing to lack of sufficient number of breeding males required in the breeding region. Moreover instead of local bull semen, exotic (Jersey and Holstein Friesian) or crossbred semen was available in the habitat. This can be one of the causes for dilution of the populations. Together these two factors were resulting in the reduction of true to the breed type animals. Actual picture of Bachaur cattle in its habitat also indicated towards the inbreeding in the population. Very less number of breeding bulls in the habitat of Bachaur cattle has been reported (Singh 2004). Bachaur is a draft purpose breed thus most of the males are used in carrying loads and agricultural operations. These males are castrated at the age of one year leading to their genetic death. Similarly over the last few years the population of Siri cattle has been declining due to extensive cross breeding with the result that Siri animals are now confined only to the remote and inaccessible areas of Sikkim. Thus, few Siri breeding bulls have been left in the habitat which might decrease the effective population size.

**Table 3** Heterozygote deficiency (F) of five cattle populations of India

| Locus | Bachaur | Gangatiri | Shahabadi | Purnea | Siri |
|---|---|---|---|---|---|
| BM1824 | 0.074 | −0.131 | 0.106 | 0.284 | 0.085 |
| CSSM08 | 0.034 | −0.102 | 0.090 | 0.007 | 0.092 |
| CSSM33 | −0.027 | −0.007 | −0.027 | 0.036 | 0.001 |
| CSSM66 | 0.072 | −0.086 | −0.063 | 0.073 | 0.027 |
| ETH10 | −0.121 | −0.124 | −0.080 | −0.424 | −0.036 |
| ETH225 | −0.015 | 0.010 | −0.139 | −0.013 | 0.067 |
| ETH3 | −0.366 | −0.182 | −0.132 | −0.316 | −0.329 |
| HEL09 | −0.129 | −0.063 | −0.036 | 0.048 | 0.128 |
| HEL5 | 0.073 | −0.060 | −0.179 | −0.135 | 0.115 |
| ILSTS06 | 0.257 | 0.151 | 0.217 | 0.213 | 0.370 |
| ILSTS11 | 0.166 | 0.188 | 0.182 | 0.123 | 0.163 |
| ILSTS34 | −0.011 | 0.089 | 0.040 | 0.134 | 0.005 |
| ILSTS33 | −0.005 | 0.047 | 0.042 | 0.184 | 0.044 |
| INRA05 | 0.170 | −0.062 | 0.090 | 0.240 | 0.023 |
| INRA35 | 0.085 | 0.147 | 0.047 | 0.081 | −0.121 |
| INRA63 | 0.134 | 0.064 | 0.122 | −0.179 | 0.098 |
| MM12 | 0.057 | 0.051 | 0.050 | 0.011 | 0.068 |
| MM8 | 0.118 | −0.088 | 0.073 | 0.040 | 0.034 |
| TGLA122 | −0.017 | −0.025 | 0.119 | 0.028 | −0.030 |
| TGLA227 | −0.153 | 0.009 | 0.082 | 0.351 | −0.023 |
| TGLA53 | −0.047 | −0.028 | 0.119 | 0.104 | 0.050 |
| Mean | **0.017** | **−0.010** | **0.034** | **0.042** | **0.040** |
| S. D. | **0.030** | **0.022** | **0.023** | **0.041** | **0.028** |

Results of F-statistics for each of the 21 loci across populations are presented in Table 4. The global deficit of heterozygotes across populations ($F_{IT}$) amounted to 7.3% ($P < 0.001$). An overall significant deficit of heterozygotes ($F_{IS}$) of 2.6% occurred in the analyzed loci because of inbreeding within populations. All loci except six contributed to heterozygote deficit within populations. The studied populations showed a low but significant genetic differentiation among five populations ($F_{ST} = 0.048$). All loci contributed to the differentiation with the highest values found for ETH25 (19.6%). Genetic differentiation of similar magnitude has been reported among cattle breeds of Orissa and hill cattle of Kumaun (0.044) from India (Sharma *et al.*2012a). Much higher $F_{ST}$ value have been reported in other indigenous cattle (Sodhi *et al.*2011). While, several reports on exotic cattle viz. North European breeds $F_{ST}$ =0.107 (McHugh *et al.*1998), seven European cattle breeds $F_{ST}$ =0.112 (Kantanen *et al.*2000) and Swiss cattle $F_{ST}$ = 0.090 (Schmid *et al.*1999) also depicted higher genetic differentiation than populations investigated in this study. The low value of genetic differentiation in Eastern Indian cattle may be attributed to the lack of high selection pressure as compared to established breeds of India or cattle breeds of developed countries. These results reflect that within-breed genetic variations is more than between-breed and this variation could be a valuable tool for genetic improvement and conservation of cattle populations of eastern India.

Pair-wise genetic differentiations quantified by $F_{ST}$ estimates identified Bachaur and Gangatiri (0.008) as the closest populations while the most differentiated were Purnea and Siri (0.044) (Table 5). Similar results were obtained with Nei's genetic distance matrix. The highest genetic distance was found between Purnea and Siri (0.292), while Bachaur and Gangatiri were closest to each other (0.037) (Figure 2). In accordance to these observations highest gene flow was among the Bachaur and Gangatiri ($N_m$ = 32.809) populations and least between Purnea and Siri ($N_m$ = 5.408). Overall $N_m$ value also indicated high rate of genetic flow between the populations (10.230 ± 1.922) (Table 4). Visualization of breed relationship was done by constructing Neighbor joining tree on the basis of Nei's genetic distance. As expected, the Siri was most distinct and separated first, while remaining populations formed two groups with clustering of Gangatiri and Bachaur on one node and Purnea and Shahabadi on second with more than 95% bootstrap value (Figure 2). This grouping pattern was further supported by PCA analysis. Since phylogenetic reconstruction may not take into account the effects of admixture between breeds, PCA was performed to further investigate possible genetic relationship between recognized breeds and local populations. First three dimension of the PCA (PC1 = 40.35; PC2 = 30.72; PC3 = 22.74) accounted for 93.81% of total variation. In the multivariate space defined by PCA, Bachaur and Gangatiri were much closer. It confirmed distinctiveness of Siri as well as lesser known cattle populations of Purnea, and Shahabadi (Figure 3).

**Table 4** Global F-statistics and estimates of Nm across five cattle populations

| Locus | Fis | Fit | Fst | Nm |
|---|---|---|---|---|
| BM1824 | 0.083 | 0.126 | 0.047 | 5.025 |
| CSSM08 | 0.025 | 0.094 | 0.070 | 3.303 |
| CSSM33 | −0.005 | 0.033 | 0.037 | 6.438 |
| CSSM66 | 0.003 | 0.014 | 0.011 | 21.733 |
| ETH10 | −0.144 | −0.088 | 0.049 | 4.834 |
| ETH225 | −0.011 | 0.188 | 0.196 | 1.023 |
| ETH3 | −0.261 | −0.208 | 0.042 | 5.726 |
| HEL09 | −0.010 | 0.017 | 0.027 | 9.078 |
| HEL5 | −0.036 | 0.053 | 0.086 | 2.648 |
| ILSTS06 | 0.242 | 0.256 | 0.017 | 14.078 |
| ILSTS11 | 0.166 | 0.229 | 0.075 | 3.091 |
| ILSTS34 | 0.049 | 0.097 | 0.051 | 4.697 |
| ILSTS33 | 0.060 | 0.119 | 0.063 | 3.732 |
| INRA05 | 0.092 | 0.098 | 0.007 | 36.707 |
| INRA35 | 0.048 | 0.159 | 0.116 | 1.905 |
| INRA63 | 0.040 | 0.056 | 0.017 | 14.799 |
| MM12 | 0.046 | 0.065 | 0.019 | 12.821 |
| MM8 | 0.036 | 0.048 | 0.012 | 21.263 |
| TGLA122 | 0.015 | 0.033 | 0.018 | 13.656 |
| TGLA227 | 0.061 | 0.084 | 0.025 | 9.857 |
| TGLA53 | 0.040 | 0.053 | 0.013 | 18.427 |
| Mean | 0.026 | 0.073 | 0.048 | 10.230 |
| SE | 0.022 | 0.022 | 0.010 | 1.922 |

**Table 5** Pair wise population matrix of Fst values between analyzed populations

| | Bachaur | Gangatiri | Shahabadi | Purnia | Siri |
|---|---|---|---|---|---|
| Bachaur | 0.000 | | | | |
| Gangatiri | 0.008 | 0.000 | | | |
| Shahabadi | 0.032 | 0.031 | 0.000 | | |
| Purnia | 0.032 | 0.033 | 0.033 | 0.000 | |
| Siri | 0.027 | 0.024 | 0.035 | 0.044 | 0.000 |

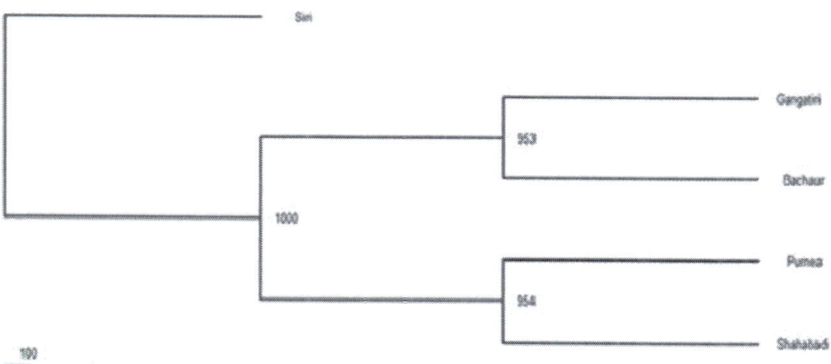


| | Bachaur | Gangatiri | Shahabadi | Purnia | Siri |
|---|---|---|---|---|---|
| Bachaur | 0.000 | | | | |
| Gangatiri | 0.037 | 0.000 | | | |
| Shahabadi | 0.191 | 0.183 | 0.000 | | |
| Purnia | 0.189 | 0.197 | 0.203 | 0.000 | |
| Siri | 0.154 | 0.136 | 0.239 | 0.292 | 0.000 |

**Figure 2** Genetic relationship among cattle populations using Nei's distance. The number in the branch indicates the percentage occurrence in 1000 bootstrap replicates. Table shows Nei's genetic distance between populations and breeds analyzed.

Among methods not assuming predefined structure, tree-based methods use genetic distance between individuals and tree construction algorithm such as UPGMA or Neighbour-joining to group them in clusters. Similarly, multivariate analyses can help in defining clusters of individuals. However these graphical methods are loosely connected to statistical procedures allowing the identification of homogeneous clusters of individuals. An alternative approach to delineate clusters of individuals on the basis of their genotypes at multiple loci was also performed using a Bayesian approach employed in software STRUCTURE. It works by grouping individuals into clusters (K) such that Hardy-Weinberg equilibrium is maximized within clusters. Likely value of K which best captured the variation present in the data was four based on modal value of K versus ΔK distribution (Figure 4) following Evano et al. (2005). Siri, Purnea and Shahabadi were grouped in their own clusters. However, Bachaur and Gangatiri animals were intermingled in one cluster (Figure 4). The results are coincident with genetic distance among the populations as divergence was lowest between Bachaur and Gangatiri. It is important to mention that geographically habitat of Gangatiri and Shahabadi are much closer than Bachaur and Gangatiri. Closeness of Bachaur and Gangatiri instead of Gangatiri and Shahabadi may be due to attempt in field to upgrade local populations with the use of semen of exotic (Jersey and Holstein Friesian) and indigenous milch (Hariana, Sahiwal, Red Sindhi, Tharparkar) cattle breeds. Secondly the habitat of Gangatiri and Shahabadi are separated by river Ganges, which may act as a geographic barrier. Based on present study Siri cattle appeared to be distinct from all other neighboring breeds and populations of eastern region of India commensurating geographical distance of its habitat resulting in reproductive isolation by distance. As per literature Bhutan is said to be the real home of this breed (Nivsarkar *et al.*2000). It was distributed from that area to the various parts of Sikkim and Darjeeling region of West Bengal states. Small cattle with similar black and white markings have been found in Sikong Province of China, which occupies a portion of the Tibetan highlands northeast of Bhutan.

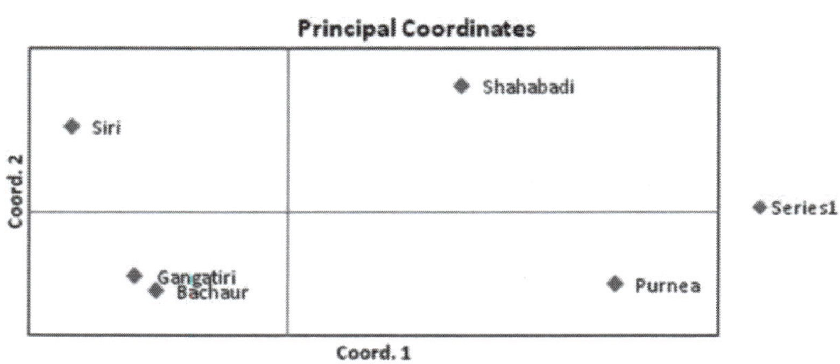


**Figure 3** Principal Coordinates Analysis (PCA) via Covariance matrix with data standardization.

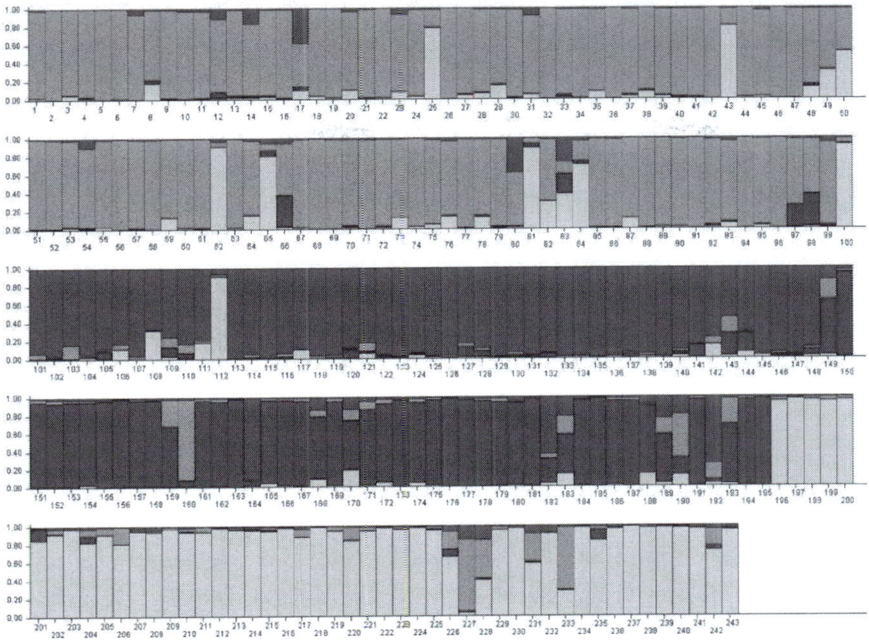

**Figure 4** Clustering assignment of 243 animals representing five East Indian cattle populations using STRUCTURE at K = 4 arranged by Q values. Each individual cattle is represented on the graph by a vertical bar divided into K colored segments corresponding to K genetic clusters. The length of each colored segment is proportional to the individual's membership in the cluster of corresponding colour. Shahabadi (Red), Purnea (Blue) and Siri cattle (Yellow) form separate cluster. Bachaur and Gangatiri (Green) cluster in one group.

# 4. CONCLUSIONS

It is vital to report that lesser known cattle populations too have high genetic diversity. Unfortunately, animals of these populations are continuously decreasing due to change in agricultural practices as mechanization is fast replacing the traditional practices involving these animals, procedures performed by breeders to increase efficiency (crossing with available exotic or crossbred semen) and apathy of government agencies since these populations are not recognized as distinct breeds. The present study established the uniqueness of two such populations; Purnea and Shahabadi from the recognized cattle breeds of Eastern India. Conservation of genetic variation in these populations should be considered by breeders, in the interest of long term future of the populations in their native tract. To begin with, breed societies need to be created, that will be responsible for registration of these cattle populations as breeds, complete maintenance and improvement of the breed to make it economically sustainable in the transforming agricultural scenario of the country.

## ACKNOWLEDGEMENTS

We acknowledge the financial Support provided by Network Project on Animal Genetic Resources, Indian Council of Agricultural Research, India for carrying out this study. The assistance and contribution of State Animal Husbandry Department, Government of Sikkim, Government of Bihar and Government of Uttar Pradesh in the collection of blood samples from habitat of local cattle is duly acknowledged.

## COMPETING INTERESTS

The authors declare that they have no competing interests.

## AUTHORS' CONTRIBUTIONS

RS participated in project conception, analysis and drafting the manuscript. AM carried out genotyping. PKS planned and collected the samples. MST helped in data analysis and writing final version of manuscript. All authors read and approved the final manuscript.

# REFERENCES


1 BAHS-Basic Animal Husbandry Statistics: *Department of Animal Husbandry, Dairying & Fisheries, Ministry of Agriculture, Government of India*. New Delhi: KrishiBhavan; 2012.

2 Bataillon TM, David JL, Schoen DJ: Neutral genetic markers and conservation genetics: simulated germplasm collections. *Genetics* 1996, 144: 409-417.

3 Buchanan FC, Adams LJ, Littlejohn RP, Maddox JF, Crawford AM: Determination of evolutionary relationships among sheep breeds using microsatellites. *Genomics* 1994, 22: 397-403. 10.1006/ geno. 1994.1401

4 Dalvit C, De Marchi M, Dal ZR, Zanetti E, Meuwissen T, Cassandro M: Genetic characterization of Burlina cattle breed using microsatellite markers. *J Anim Breed Genet* 2008, 125: 137-144. 10.1111/j.1439-0388.2007.00707.x

5 Dodgson JB, Cheng HH, Okimoto R: DNA marker technology: a revolution in animal genetics. *Poultry Sci* 1997, 76: 1108-1114.

6 Earl DA, vonHoldt BM: STRUCTURE HARVESTER: a website and program for visualizing STRUCTURE output and implementing the Evanno method. *Conserv Genet Resour* 2012, 4(2):359-361. 10.1007/s12686-011-9548-7

7 Egito AA, Paiva SR, Albuquerque MM, Mariante AS, Almeida LD, Castro SR, Grattapaglia D: Microsatellite based genetic diversity and relationships among ten Creole and commercial cattle breeds raised in Brazil. *BMC Genet* 2007, 8 83. 10.1186/1471-2156-8-83

8 Evanno G, Regnaut S, Goudet J: Detecting the number of clusters of individuals using the software structure: A simulation study. *Mol Ecol*2005, 14: 2611-2620. 10.1111/j.1365-294X.2005.02553.x

9 FAO: *The state of the world's animal genetic resources for food and agriculture*. Edited by: Rischkowsky B, Pilling D. Rome, Italy: FAO; 2007.

10 FAO [Food and Agricultural Organization of the United Nations]: *Food and Agriculture Organization.* 2011. Molecular genetic characterization of animal genetic resources. FAO Animal Productio and Health Guidelines. No. 9. Rome, repost.

11 Felsenstein J: *PHYLIP: phylogeny inference pakage, version 3.5.* Seattle, Washington: Department of Genome Sciences, Washington University; 1993.

12 Joshi BK, Sodhi M, Mukesh M, Mishra BP: Genetic characterization of farm animal genetic resources of India: A review. *Indian J of Anim Sci* 2012, 82(3):1259-1275.

13 Kantanen J, Olsaker I, Holm LE, Lien S, Vilkki J, Brusgaard K, Eythorsdottir E, Danell B, Adalsteinsson S: Genetic diversity and population structure of 20 North European cattle breeds. *J Hered* 2000, 91: 446-457. 10.1093/jhered/91.6.446

14 McHugh DE, Loftus RT, Cunningham P, Bradley DG: Genetic structure of seven European cattle breeds assessed using 20 microsatellite markers. *Anim Genet* 1998, 28: 333-340.

15 Metta M, Kanginakudru S, Gudiseva N, Nagaraju J: Genetic characterization of the Indian cattle breeds, Ongole and Deoni ( *Bos indicus* ), using microsatellite markers – a preliminary study. *BMC Genet* 2004, 5: 16.

16 Mukesh M, Sodhi M, Bhatia S, Mishra BP: Genetic diversity of Indian native cattle breeds as analyzed with 20 microsatellite loci. *J Anim Breed Genet* 2004, 121: 416-424. 10.1111/j.1439-0388. 2004.00468.x

17 Nei M: *Molecular evolutionary genetics*. New York USA: Columbia University Press; 1987.

18 Nivsarkar AE, Vij PK, Tantia MS: *Animal genetic resources of India: Cattle and buffalo*. India: ICAR; 2000.

19 Page RDM: TreeView: An application to display phylogenetic trees on personal computers. *Comp App in Biosci* 1996, 12: 357-358.

20 Pandey AK, Sharma R, Singh Y, Prakash B: Genetic diversity studies of Kherigarh cattle based on microsatellite markers. *J Genet* 2006, 85: 117-122. 10.1007/BF02729017

21 Pandey AK, Sharma R, Singh Y, Prakash B, Ahlawat SPS: Evaluation of Genetic Variation in Kenkatha cattle by microsatellite markers.*Asian–Aust J Anim Sci* 2006, 19: 1685-1690.

22 Peakall R, Smouse PE: A heterogeneity test for fine-scale genetic structure. *Mol. Ecol. Notes* 2008, 17: 3389-3400. 10.1111/j.1365-294X.2008.03839.x

23 Petit RJ, EI Mousadik A, Pons O: Identifying populations for conservation on the basis of genetic markers. *Conserv Biol* 1998, 12: 844-855. 10.1046/j.1523-1739.1998.96489.x

24 Pritchard J, Stephensa M, Donnelly P: Inference of population structure using multilocus genotype data. *Genetics* 2000, 155: 945-959.

25 Sambrook J, Fritsch EF, Maniatis T: *Molecular cloning: a laboratory manual*. Cold Spring Harbour, NY: Cold Spring Harbour Lab. Press; 1989.

26 Schmid M, Saipveaova N, Gaillard C, Dolf G: Genetic diversity in Swiss cattle breeds. *J Anim Breed Genet* 1999, 116: 1-8. 10.1111/j.1439-0388.1999.00165.x

27 Sharma R, Pandey AK, Singh SCP, Singh Y, Mishra BP, Singh SR, Mandal KG, Singh PK, Singh G, Ahlawat SPS: Diversity assessment of Bachaur cattle from India: A draft breed still revelant in mechanized Era. *Korean J Genet* 2007, 29: 369-378.

28 Sharma R, Pandey AK, Singh Y, Prakash B, Mishra BP: Estimation of genetic diversity in Siri cattlefrom India. *Russian J Genet* 2008, 44(11):1532-1538.

29 Sharma R, Pandey AK, Singh Y, Prakash B, Mishra BP, Kathiravan P, Singh PK, Singh G: Evaluation of genetic variation and phylogenetic relationship among north Indian cattle breeds. *Asian–Aust J Anim Sci* 2009, 22: 13-19.

30 Sharma R, Maitra A, Pandey AK: Genetic structure and differentiation of four Indian autochthonous cattle populations. *Russian J Genet*2012, 48(6):611-617. 10.1134/S1022795412060129

31 Sharma R, Pandey AK, Singh PK, Maitra A, Mukesh M, Singh SR, Singh B: Characterization of Shahabadi cattle of Bihar-Phynotypic and molecular approaches. *Indian J Anim Sci* 2012, 82(3):318-322.

32 Sharma R, Singh PK, Maitra A, Pandey AK, Mukesh M, Singh SR, Singh B: Molecular characterization, body parameters and management practices of Purnea cattle. *Indian J Anim Sci* 2013, 83(5):536-541.

33 Singh SR: *Project report on survey and characterization of Bachaur cattle in Bihar submitted to ICAR*. Karnal: New Delhi through NBAGR; 2004.

34 Singh SR: *Conservation of threatened breeds of livestock and poultry*. Patna, India: Bihar Agricultural Management and Extension Training Institute; 2009.

35 Singh PK, Gaur GK, Pundir RK, Singh A: Characterization and evaluation of Gangatiri cattle breed in its native tract. *Indian J Anim Sci*2007, 77: 66-70.

36 Sodhi M, Mukesh M, Ahlawat SPS, Sobti RC, Gehlot GC, Mehta SC, Prakash B, Mishra BP: Genetic diversity and structure of two prominent Zebu cattle breeds adapted to the arid region of India inferred from microsatellite polymorphism. *Biochem Genet* 2008, 46: 124-136. 10.1007/s10528-007-9135-5

37 Sodhi M, Mukesh M, Mishra BP, Ahlawat SPS, Prakash B, Sobti RC: Microsatellite analysis of genetic population structure of Zebu cattle (*Bos indicus* ) breeds from North-Western region of India. *Anim. Biotech* 2011, 22: 16-29. 10.1080/10495398.2011.536091

38 Sun W, Chen H, Lei C: Genetic variation in eight Chinese cattle breeds based on the analysis of microsatellite markers. *Genet Sel Evol*2008, 40: 681-692. 10.1186/1297-9686-40-6-681

# CHAPTER 5

## Phenotypic Characterization and Strategies for Genetic Improvement of WAD Goats under Backyard Systems

**S. O. Oseni[1*], B. A. Ajayi[2]**

[1]Department of Animal Sciences, Obafemi Awolowo University, Ile-Ife, Nigeria

[2]Department of Animal Science, Landmark University, Omu-Aran, Nigeria

## ABSTRACT

West African Dwarf (WAD) goats represent one of the predominant small ruminant breeds raised by resource-limited households in southern Nigeria as well as in the humid west and central Africa. This breed of goats is renowned for its adaptation to hot and humid environments, high fertility and prolificacy under backyard systems where they are raised almost with zero investment. On account of its contributions to household income and food security in southern Nigeria, there is the need for a detailed phenotypic and genetic characterization as well as the design of breeding strategies for its conservation through sustainable utilization. A critical challenge, however, is that there are no national breeding policies in most of the countries where WAD goats are raised. The objectives of this paper are: to review literature reports on trait-level information and estimates of genetic parameters for some economic characters of WAD goats; and to discuss a communal breeding scheme for its conservation through sustainable utilization. The presentation shows that WAD goats represent a highly diverse population, with particular reference to qualitative and performance-related traits, reflecting the role of traditional institutions in the maintenance of animal diversity in local populations. A genetic improvement programme involving dispersed open nucleus breeding scheme was discussed. The scheme involves backyard WAD goat units in villages as base population and dispersed communally managed schemes that maintain a population of bucks only, as nucleus units. Critical issues such as the definition of breeding goals by farmers, as well as communal project ownership, are emphasized. The ultimate goal of the breed- ing programme is to create opportunities for community-based management of WAD goat genetic resources for the benefit of backyard goat farmers in the region.

## KEYWORDS

WAD Goats, Characterization, Trait-Level Information, Genetic Improvement, Communal Breeding Scheme, Backyard Systems

## 1. INTRODUCTION

West African Dwarf goats (also called Nigerian Dwarf, Kirdi or Kirdimi, Fouta Djallon, Cameroon Dwarf, Af- rican Pigmy [1] [2] are thought to have evolved from the Brevipes or short-legged goats. Epstein [3] described WAD goats as those whose average shoulder height does not exceed 50 cm. Years of adaptation and natural se- lection under humid tropical conditions made this breed highly adapted to the humid forest zone [4]. These goats are present in all of humid Africa, from Southern Sudan to the west coast areas that are noticeably humid and warm [1] , and characterized by dense vegetation ranging from swamp mangrove to rainforest and derived sa- vannah. In these zones, ambient temperature and relative humidity are notably high all year round. Gall [1] and DAGRIS [2] listed some 16 countries in West and Central Africa where WAD goats are predominantly raised under backyard systems. The production system is mainly extensive, animals are allowed to roam and scavenge at will, with a preponderance of subsistence farming units. These countries are noted for high poverty levels [5] , especially in the rural areas. Among the inhabitants, WAD goats are raised principally as a source of meat and other uses include socio-economic, ceremonial, cultural and traditional purposes, including divination.

WAD goats are highly prolific, can be bred all year round, with up to three parturitions in two years [6] . The goats are hardy, with the ability to thrive and survive under harsh environmental conditions of heat and humidity, ability to digest a broad range of diets and resistance to high-humidity pathogens and haemonchosis, tolerant of gastro-intestinal nematodes and trypanosomiasis [7] - [9] and the ability to thrive in tse-tse fly infested humid forests and guinea savannah zones [10] [11] . [12] noted that the mechanism of trypano-tolerance in goats, in general, differs from that of trypano-tolerant cattle. WAD goats have been described as trypano-“resilient”, since, in the face of parasitic challenge, they maintained positive weight gain and recorded low mortality rates. This allows them to graze on land not available to other domestic livestock [8] . This author noted further that the ability of WAD goats to survive under adverse environmental conditions with low inputs makes this breed a low-risk choice, and this, according to [12] , indicates their potential role for a more sustainable solution to the growing demand for animal protein in developing countries. Thus, [13] described WAD goats as an extremely valuable genetic resource on account of their entire package of adaptive traits.

Several investigators [1] [10] [13] [14] have reported cases of indiscriminate mating of WAD goats with Sa- helian breeds of goats. In Nigeria, [10] noted specific cases of mating of WAD goats with the Red Sokoto breed, particularly in the transition zone between the rainforest and the derived savannah belt. In Burkina Faso, [14] reported that gene flow among Burkinabe goat populations (Djallonké or WAD goats, Mossi and Sahelian) was very high and this accounted for the low genetic differentiation among these breeds. All these practices contribute to an admixture of the gene pool which the WAD goat represents, indicating a general threat of introgression and dilution of its genetic base.

On account of its contributions to household income and food security in southern Nigeria, there is the need for a detailed genetic characterization as well as the design of breeding strategies for its conservation through sustainable utilization. A critical challenge however, is that there are no national breeding policies in most of the countries where WAD goats are raised, in spite of their contributions to the diversified livelihoods of resource- limited farmers and to the rural economies. The objectives of this paper are: 1) To review literature reports on trait-level information (e.g. qualitative, morphological, functional and performance-related traits) as well as es- timates of genetic parameters for these traits under on-station and on-farm conditions; 2) To discuss a communal breeding scheme for the conservation of WAD goats through sustainable utilization under backyard production systems in the humid west and central Africa.

## 2. TRAIT-LEVEL INFORMATION

### 2.1. Qualitative Traits

WAD goats display wide variations in qualitative traits including coat colour pigmentation, wattles, beard and super-numerary teats among females. Coat colour pigmentations include predominantly black, brown white, pied and mixed colours [15] [16] . Figure 1 shows pictures of WAD goats from Ile-Ife and its environs in south-western Nigeria, displaying all the afore-mentioned coat colours under backyard systems in the region. [15] noted that WAD goats had a preponderance of basic black coat colour (53.3%) while basic brown and white goats were 39.9 and 6.8% respectively. Other qualitative traits displayed by WAD goats include wattles (bilat- eral, unilateral and WAD goats with no wattles), with frequency distributions of about 63%, 6.2% and 30.9%, respectively, while the possession of beard is common among all breeding males and about 12% of breeding does.

**Figure 1**. Diversity in coat colours of WAD goats under backyard systems in south-western Nigeria (Courtesy: S Oseni, WAD Goat field studies, southwest Nigeria, 2008).

Results of some fixed model analysis revealed that there were no significant differences among coat colour classes of dams in prolificacy, litter size at weaning and birth weight [17] . This author however, reported that WAD goats with basic white colours were significantly heavier at weaning and at yearling ages than basic brown and black goats. This observation was supported by [16] who reported that body weight of WAD goats decreased with increasing coat colour pigmentation. Other results of fixed model analysis showed that the possession of wattles did not exert any influence on the traits studied except yearling weight where bilaterally-wattl- ed WAD goats were lighter in weight [17] . According to [18] , visually obvious features such as coat colours and horns in primitive breeds of sheep are linked to properties of constitution that affect animal survival. Large scale studies are however, needed to establish the true roles of these qualitative traits on animal adaptation and per- formance and overall productivity of WAD goats under backyard systems.

## 2.2. Reproductive Traits

These include fertility, prolificacy, annual doe productivity and kid survival rate of WAD goats are shown in Table 1. Key literature sources for these information include the detailed review of goat breeds of Nigeria by [10] and scientific papers by [19] - [24] , as well as the classical treatise “Goat Breeds of the World” [1] .

According to [23], majority (75%) of female WAD goats had their first kids at ages ranging between 14 and 20 months, which agreed with an earlier report [10] that the overall mean of age at first kidding was 18.5 ± 1.2 months. Doe prolificacy was 1.70 ± 0.04 [20] and was markedly affected by year of birth and parity [20] [25] . The relationship between litter size at birth and parity shows that prolificacy rose with parity up the $6^{th}$, and re- mained stable thereafter. With increasing parity, does are more mature and efficient in reproduction [25] . Over- all means for kidding interval were 230 ± 67 to 283.15 ± 6.79 days [10] [21] . This trait was significantly af- fected by parity, year and season of birth and litter size at birth [20] . The relationship between kidding interval and doe parity showed that prolonged kidding intervals were associated with the first three parities, after which kidding intervals remained relatively constant at about 250 d [20] . Frequency distributions of kidding interval [23] revealed that almost 50% of the does had kidding interval ranging between 200 and 250 d while about one-third and one-fifth of the does had parturition intervals of over 250 d and less than 200 days, respectively. With prolonged kidding intervals, the goal of having three parturitions in two years becomes increasingly diffi- cult to achieve.

Pre-weaning mortality rates (from 1 to 3 months of age) and annual mortality rates were 22.6% ± 15.9% and 5%, respectively, [23] [24] and could be as high as 50% [6] . Incidence of high kid mortality reduces doe produc- tivity, as well as the contributions of WAD goat production to sustainable livelihoods. Survival rate of kids was also markedly affected by type of birth, with a higher survival rate associated with kids in single births when compared to triplets and quadruplets [20] .

Annual doe reproductive index (Kg weaned offspring per doe per year) and doe productivity index (Kg year- lings per doe per year [24] ) used to assess the overall performance of WAD goat does under on-station and vil- lage conditions are also presented in Table 1. These indices combine fertility, prolificacy, kid survival to wean- ing and yearling ages, weight at weaning and yearling ages as well as kidding interval. Means for reproductive index ranged from 5.7 to 9.88 [26] [27] . Doe productivity index ranged from 33.3 ± 14.3 to 37.1 ± 18.1 [24] . According to [10] , WAD goats under restricted environments have not generally outperformed their free ranging counterparts, suggesting that there may still be some management issues that are not well understood or applied.

**Table 1**. Trait-level information from reproductive, growth and morphological characteristics of WAD goats from the litera- ture.

| Trait | n | Mean (±SE) | Source |
|---|---|---|---|
| **i. Reproductive and doe productivity index** | | | |
| Age at first kidding (months) | 371 | 18.5 ± 1.2 (SD)[1] | [10] |
| Litter size at birth | 220 | 1.70 ± 0.04 | [20] |
| Kidding interval (days) | 209 | 283.15 ± 6.79 | [20] |
| Kidding interval (days) | ----- | 230 ± 67 to 274 ± 75 | [21] |
| Mortality (1 - 3 months, %) | ----- | 22.6 ± 15.9 | [22] |
| Pre-weaning mortality (% born) | ----- | 17.3 to 28.8 | [24] |
| Mortality to 1 year (% of number weaned) | ----- | 0 to 4.9 | [24] |
| Annual mortality rate (%) | ----- | 29.1% to 49.5% (range) | [21] |
| Reproductive index (Kg) | 245 | 6.5 ± 3.7 | [22] |
| Reproductive index (Kg weaned/doe/year) | 33 | 6.5 ± 5.5 to 22.0 ± 6.6 | [24] |
| Reproductive index (Kg weaned/doe/year) | ---- | 9.88 | [20] |
| Annual doe productivity index (Kg yearlings/doe/year) | ---- | 5.3 ± 3.9 to 9.0 ± 7.2 | [21] |
| Productivity index (Kg yearlings/doe/year) | 33 | 33.3 ± 14.3 to 37.1 ± 18.1 | [24] |
| **ii. Live body weight (Kg)** | | | |
| Kid birth weight | 62 | 1.40 ± 0.30 | [24] |
| Weaning weight | 49 | 6.40 ± 1.0 | [24] |
| Body weight at 5 months | ----- | 5.95 ± 0.14 | [19] |
| Live weight at 13 weeks | 1045 | 4.70 ± 1.3 | [22] |
| Live weight at 26 weeks | 654 | 6.1 ± 1.8 | [22] |
| Live weight at 39 weeks | 365 | 7.3 ± 2.2 | [22] |
| Live weight at 52 weeks | 219 | 8.4 ± 2.6 | [22] |
| Body weight at 12 months | ---- | 9.49 ± 0.28 | [19] |
| Mature body weight (Kg) | ---- | 20 to 30 | [6] |
| **iii. Morphmetric characteristics** | | | |
| Height at withers (male, cm) | ---- | 46.0 ± 1.2 (SD) | [10] |
| Height at withers (female, cm) | ---- | 44.0 ± 1.5 (SD) | [10] |
| Body length (cm) | 770 | 53.49 ± 0.41 | [16] |
| Height at withers | 770 | 43.87 ± 0.39 | [16] |
| Abdominal girth | 770 | 65.30 ± 0.87 | [16] |

[1]Standard deviation.

## 2.3. Live Body Weight

Overall means for live body weight at birth and at different ages up to yearling weight are presented in Table 1. Mean kid birth weight of 1.28 ± 0.01 Kg was significantly affected by type of birth (singletons, twins, triplets or quadruplets) and by sex of kid [23] . According to this author, singletons (whether males or females) were con- sistently heavier at birth than kids in twins, triplets and quadruplets. Similarly, among all litter types, males were heavier than females. These authors noted that the difference between sexes was however, more manifest in sin- gletons. Further, body weight from birth to $9^{th}$ week of kids derived from different litter types revealed that kids in singletons were heaviest at all ages from birth to $9^{th}$ week of age, while kids from quadruplets weighed the least [23].

## 2.4. Morphometric Characters

The morphometric traits of adult WAD goats are presented in Table 1. These variables include body length, body depth, height at withers and rump [16]. Linear measurements, like live body weight, showed variations among adult WAD goats, suggesting that this breed of goats is largely unselected. Among the morphological variables, heart girth had the highest correlation ($r = 0.88$; $P < 0.001$) with live body weight (Oseni and Ajayi, 2014, unpublished). These authors also observed that heart girth, whether in linear or multiple regression models accounted for over 77% of total variability in live weight among WAD goats and showed that the best fit predic- tion model for live weight of WAD goats included heart girth and one or both of body depth and body length. Similarly, [28] showed that the coefficient of determination of multiple regression models including heart girth and any other linear body measurement on body weight was slightly higher than that of simple regression of heart girth on body weight. Further studies [29] focussed on the comparative multivariate analysis of biometric traits of WAD and Red Sokoto goats. For WAD goats, the study extracted four principal components (PC) with the following loadings: the first PC which had its loadings for chest depth, chest circumference, abdominal cir- cumference, body weight and neck circumference accounted for 66.29%

of total variance. The second PC which was dominated by foreleg length, hindleg length, withers height and rump height covered 9.37% of total vari- ance. The third PC had its loadings on rump length, rump width, head width, shoulder width and body length accounted for 5.48% of total variance. The final PC which was only dominated by horn length covered 4.73% of total variance. The study showed clear spatial racial separation and distinctness in morphology between WAD and Red Sokoto goats.

## 3. GENETIC PARAMETERS

Table 2 shows estimates of heritability for some economic traits of WAD goat populations from the literature [30] - [33], with the most recent and comprehensive estimates of genetic parameters by [11] . Literature on esti- mates of genetic parameters are rather few, indicating that research work on WAD goat breeding and genetics were scarce, suggesting a low level of priority for this area of research. In general, all reproductive traits (in- cluding litter size at birth and kidding interval) were low to moderately heritable, as expected for traits related to reproductive and functional traits, on account of the high environmental effects on these traits. Traits that were highly heritable included kid birth weight, sex ratio and sex combination (among twins and triplets) as reported by [32] . These estimates were mostly derived from small sample sizes and/or accompanied by large standard errors, with the exception of [11] , which is worthy of note because of the exceptionally large sample sizes and the fact that animals were raised under low trypano-tolerance challenge in low-input production environment.

Genetic correlations [11] [34] among body weights at weaning and at different ages up to yearling weight were positive and significant. In particular, [34] noted that the genetic correlations between weaning weight and three-month body weight ($r = 0.65$; $P < 0.01$) was the highest, and concluded that selection for body weight at early ages (especially at three months) will be beneficial in improving this trait at subsequent ages. Similarly, [11] noted that genetic correlations were mostly moderate to high and suggested that substantial improvement can be made by selection.

**Table 2**. Heritability estimates for some economic traits in WAD goats from literature.

| Trait | Heritability (±SE) | Source |
|---|---|---|
| Birth weight | 0.50 ± 0.05 | [11] |
| Birth weight | 0.59 ± 0.45 | [32] |
| Litter size at birth | 0.32 ± 0.07 | [33] |
| Litter size at birth | 0.08 to 0.18 (range) | [30] |
| Kidding interval | 0.03 ± 0.01 | [33] |
| Weaning weight | 0.14 ± 0.003 | [31] |
| Weaning weight | 0.43 ± 0.07 | [11] |
| 3-Month body weight | 0.29 ± 0.005 | [31] |
| Pre-weaning growth rate | 0.32 ± 0.08 | [11] |
| 6-Month body weight | 0.11 ± 0.003 | [31] |
| Post-weaning growth rate | 0.11 ± 0.05 | [11] |
| 9-Month body weight | 0.17 ± 0.004 | [31] |
| Yearling weight | 0.30 ± 0.07 | [11] |
| Sex ratio | 0.72 ± 0.54 | [32] |
| Sex combination (twins & triplets) | 0.69 ± 0.66 | [32] |

On genetic characterization however, it was noted that there are very few reports in the literature that provides detailed information on molecular characterization of WAD goats. [14] determined molecular co-ancestry and the degree of admixture among three goat populations including Djallonké (or WAD goats) in Burkina Faso. Detailed genetic characterization on WAD goats could be done to establish the following: number of alleles, their frequency and diversity, polymorphic information content, as well as expected and observed heterozygosity [35] . [36] , using a panel of 21 microsatellite markers, showed that the genetic distance between WAD and Ma- radi goats was 0.39, noting that the genetic diversity among their crosses could generate good heterotic advan- tage and combining abilities. Further, [37] presented data to describe the genetic diversity among the three major Nigerian goat breeds (WAD, Red Sokoto and Sahel goats) at the DQB1 immune gene locus. The study revealed the following: 1) high level of genetic polymorphism among Nigerian goat breeds; 2) the phylogenetic tree from consensus sequences placed Red Sokoto and Sahel goats closer to each other than either was to the WAD goats; 3) sequence polymorphisms among these Nigerian goat breeds may be used in the search for association with disease resistance. On account of its current risk status [2] , such comprehensive characterization studies that combine phenotypic, molecular genetic information and field data become imperative.

# 4. CONSERVATION OF WAD GOATS THROUGH UTILIZATION

## 4.1. Goals

The ultimate goals of a communal breeding scheme for WAD goats are to: 1) improve the risk status of WAD goats through conservation measures that are hinged on sustainable utilization; and 2) provide backyard WAD goat farmers with improved germ-plasm, through the supply of superior WAD goat bucks for mating.

## 4.2. Considerations

Critical considerations in the design of conservation measures for WAD goats under backyard systems include the following: 1) taking cognizance of the needs, priorities and perceptions of flock owners in the design and formulation of breeding programmes and strategies [38] ; 2) ensure that breeding programmes are compatible with the needs of goat farmers and their production system; 3) that the ensuing breeding programmes are rela- tively simple in design and involve low risks [39] ; 4) accommodate the role of local knowledge systems in the maintenance of a diverse genetic pool of WAD goats, in particular, and animal care and management in general [40] .

Thus, from the foregoing, the definition of breeding objectives will be done by the WAD goat farmers. Such breeding objectives should take into consideration, the multi-functional roles of WAD goats in the traditional setting in West and Central Africa, including the following: source of meat, ceremonies, savings (e.g. bank- on-hooves), settlement of dowries/bride prizes and other traditional roles (e.g. sacrifice and divination), where WAD goats with specific coat colours are required. According to [41] , breeding policies need to incorporate such multi-functional roles in the design of breeding programmes in order to guarantee their success. [42] out- lined detailed community-based organization of genetic improvement of livestock production while [43] pre- sented results of a participatory process of the ranking of traits of WAD goats by farmers in Benin Republic. Such participatory and community-based approaches to the design of livestock improvement schemes represent a critical step in the definition of breeding objectives and contribute to the development of sustainable commu- nal breeding schemes.

## 4.3. Challenges

Major challenges in the implementation of a communal breeding scheme for WAD goat owners include: 1) among stock owners, the practice of selling off the fastest growing males for slaughter, leaving the slow-grow- ing type in the flocks, thereby creating negative selection and reducing the quality of their flocks [44] [45] ; 2) small flock sizes of backyard WAD goat units [6] ; 3) the involvement and mobilization of farmers in the design and implementation of communal breeding schemes in terms of project ownership and as final beneficiaries [38] [46] ; 4) establishing and sustaining an on-farm recording system, a pre-requisite for any successful breeding programme [41] . Solutions to these challenges will lead to the design and implementation of sound communal breeding schemes for WAD goats under backyard systems.

## 4.4. Communal Breeding Schemes

A proposed breeding scheme for WAD goats under backyard systems can be operated on a communal or coop- erative basis, and could involve the implementation of a dispersed open nucleus breeding programme described in [47] . The choice of a communal scheme is due to the fact that the pre-dominant backyard WAD goat units found all across west and central Africa are devoid of sophisticated breeding centres involving progeny testing or artificial insemination. Thus, a cooperative breeding scheme that is relatively simple in design and imple- mentation [39] , owned and managed by the farmers becomes a viable option. The ensuing breeding scheme will contribute to the maintenance and provision of WAD goat germ-plasm for use by farmers, especially by supply- ing superior bucks for natural mating in village flocks. The use of dispersed open nucleus systems [47] is par- ticularly advantageous because it will promote animal (buck) exchange between communities, thereby help to stem the tide of inbreeding within villages, and facilitate conservation and diversity in the gene pool of WAD goats. Key steps in the actualization of the breeding plan are hinged on the following: 1) Farmer mobilization in each locality or region; 2) formation of a WAD goat cooperative breeding scheme, ultimately to be owned, run and managed by the farmers on a cooperative basis; 3) animal identification and on-farm recording. The imple- mentation of the breeding scheme could be based on the open nucleus breeding scheme that maintains males only, as reported by [48] and [49] for Djallonké sheep in Cote d'Ivoire. For simplicity and ease of implementa- tion, the screening of young bucks of WAD goats could involve two phases—1) an on-farm pre-selection phase, based on live weight at 90 d; and 2) an on-station final selection phase based on yearling weight, followed by the distribution of the selected bucks to participating farmers. Critical to the success of this communal breeding scheme is the dissemination and efficient use of superior bucks from the nucleus units, among all registered and participating farmers, through some traditional pre-agreed stock sharing formula to be designed and imple- mented by them.

## 5. CONCLUSION

The paper draws attention to current information on the risk status and characteristics of WAD goats under backyard systems in west and central Africa. Details were provided on trait-level information relating to qualita- tive, reproductive, growth and morphological characters. Few estimates of genetic parameters and molecular characterization in the literature indicate the low priority accorded breeding and genetics in WAD goat research. Critical steps in the design of a communal genetic improvement scheme for WAD goats based on a dispersed open nucleus breeding scheme were discussed. The ultimate goal is to provide WAD goat germplasm through the supply of superior WAD bucks to participating farmers in backyard systems in villages.

## ACKNOWLEDGEMENTS

1) University Research Committee, Obafemi Awolowo University, Ile-Ife, Nigeria for a research grant code: 11812 AWQ; 2) WAD Goat Project (1982 to 1993) conducted at the Obafemi Awolowo University, Ile-Ife, Ni- geria, with financial support from the government of the Netherlands; 3) The dedicated owners of backyard WAD goat production system, who constitute our clientele group.

## REFERENCES

1 Gall, C. (1996) Goat Breeds of the World. CTA—Margraf Verlag, Germany, 186 p.

2 DAGRIS (2008) Domestic Animal Genetic Resources Information System (DAGRIS). General Breed Information for Djallonke Goats. http://dagris.ilri.org/dagris/location.asp?ID=803

3 Epstein, H. (1971) The Origin of the Domestic Animals of Africa. Africana Publication Corporation, New York, 214- 220.

4 Leak, S., Faye, D. and Geerts, S. (2002) Genetic Resistance to Vectors and Vector-Borne Diseases in Indigenous West African Livestock. CPTV Newsletter, 6, 36-38.

5 Central Intelligence Agency (CIA) (2008) World Fact Book. Country Report on Nigeria and Other Countries in West and Central Africa. https://www.cia.gov/library/publications/the-world-factbook/geos/ni.html

6 Ademosun, A.A. (1993) The Scope for Improved Small Ruminant Production in the Humid Zone of West and Central Africa: The Approach of the West African Dwarf Goat Project. In: Ayeni, A.O. and Bosman, H.G., Eds., Goat Produc- tion Systems in the Humid Tropics, Pudoc Scientific Publishers, Wageningen, 2-13.

7 Devendra, C. and Mcleroy, G.B. (1982) Goat and Sheep Production in the Tropics. Intermediate Tropical Agricultural Series, Longman Scientific and Technical Publishers, Longman, London, 218-219.

8 Rege, J.E.O. (1994) Indigenous African Small Ruminant Resources: A Case for Characterization and Improvement. In: Lebbie, S.H.B., Rey, B. and Irungu, E.K., Eds., Small Research & Development in Africa. Proceedings of the 2nd Bi- ennial Congress of the African Small Ruminant Research Network, Arusha, 7-11 December 1992, 205-212.

9 Blench, R. (1999) Traditional Livestock Breeds: Geographical Distribution and Dynamics in Relation to the Ecology of West Africa. Working Paper 122, Overseas Development Institute (ODI), London.

10 Ngere, L.O., Adu, I.F. and Okubanjo, I.O. (1984) The Indigenous Goats of Nigeria. Animal Genetic Resources Infor- mation, 3, 1-9.

11 Bosso, N.A., Cissé, M.F., van der Waaij, E.H., Falla, A. and van Arendonk, J.A.M. (2007) Genetic and Phenotypic Parameters of Body Weight in West African Dwarf Goat and Djallonké Sheep. Small Ruminant Research, 67, 271-278.http://dx.doi.org/ 10.1016/ j.smallrumres.2005.11.001

12 Goosen, B. (1998) Sustainability of Small Ruminant Production in the Gambia with Special Reference to the Improved Husbandry of Trypanotolerant Sheep. Veterinary Parasitology, 79, 283-297.

13 Hoeven, E., Leak, S., Hanotte, O., Jianlin, H. and Fidalis, M.N. (2004) Diversity and Mixture in West African Goats. ITC Newsletter, January-June 2004, 5-17.

14 Traoré, A., Álvarez, I., Tambourá, H.H., Fernández, I., Kaboré, A., Royo, L.J., Gutiérrez, J.P., Sangaré, M., Ouédraogo- Sanou, G., Toguyeni, A., Sawadogo, L. and Goyache, F. (2008) Genetic Characterisation of Burkina Faso Goats Using Microsatellite Polymorphism. Livestock Science, 123, 322-328.

15 Odubote, I.K. (1992) Characterization of the WAD Goat for Certain Qualitative Traits. Nigerian Journal of Animal Production, 19, 37-41.

16 Ozoje, M.O. and Mgbere, O.O. (2002) Coat Pigmentation Effects in West African Dwarf Goats: Live Weights and Body Dimensions. Nigerian Journal of Animal Production, 29, 5-10.

17 Odubote, I.K. (1994) Influence of Qualitative Traits on the Performance of West African Dwarf Goats. Nigerian Journal of Animal Production, 21, 25-28.

18 Jewell, P.A. (1998) Behavioural and Other Aspects of the Adaptations of Rare Breeds to Their Surroundings. In: Lewis, R.M., Alderson, G.L.H. and Mercer, J.T., Eds., The Potential Role of Rare Livestock Breeds in UK Farming Systems, British Society of Animal Science Meeting and Workshop Publication, 21-27.

19 Sumberg, J.E. and Mach, S.D. (1985) Village Production of WAD Goats and Sheep in Nigeria. Tropical Animal Health and Production, 17, 135-140.

20 Odubote, I.K., Akinokun, J.O. and Ademosun, A.A. (1993) Production Characteristics of West African Dwarf Goats under Improved Management Systems in the Tropics. In: Ayeni, A.O. and Bosman, H., Eds., Goat Production Systems in the Humid Tropics, Proceedings of an International Workshop, Pudoc Scientific Publishers, Wageningen, 202-207.

21 Broek van den, A. (1993) Goat Production in the Mono-Department (Republic of Benin): From Survey Methodology to Research Proposals. In: Ayeni, A.O. and Bosman, H., Eds., Goat Production Systems in the Humid Tropics, Proceedings of an International Workshop, Pudoc Scientific Publishers, Wageningen, 154-159.

22 Bosman, H.G. and Ayeni, A.O. (1993) Zootechnical Assessment of Innovations as Adapted and Adopted by the Goat Keepers. In: Ayeni, A.O. and Bosman, H.G., Eds., Goat Production Systems in the Humid Tropics, Pudoc Scientific Publishers, Wageningen, 45-57.

23 Egbunike, G.N., Akusu, M.O. and Carew, B.A.R. (1993) Vital Reproductive Statistics of West African Dwarf Does. In: Ayeni, A.O. and Bosman, H.G., Eds., Goat Production Systems in the Humid Tropics, Pudoc Scientific Publishers, Wageningen, 208-212.

24 Bosman, H.G. (1995) Productivity Assessments in Small Ruminant Improvement Programmes. A Case Study of the West African Dwarf Goat. Thesis, Grafisch Bedrijf Pons & Looijen b.v. Wageningen, Wageningen, 224 p.

25 Chiboka, O., Somade, B. and Montsma, G. (1988) Reproduction of WAD Goats-A Summary of Research at Ile-Ife, Nigeria. In: Smith, O.B. and Bosman, H.G., Eds., Goat Production in the Humid Tropics, Proceedings of a Workshop at the University of Ife, Ile-Ife, 20-24 July 1987, 125-136.

26 Reynolds, L. and Adeoye, S.A.O. (1986) Planted Leguminous Plants and Livestock Production. The International Workshop on Alley Farming for Humid and Sub-Humid Regions of Tropical Africa, International Institute for Tropical Agriculture (IITA), Ibadan, 13.

27 Adu, I.F., Adeniyi, A.O. and Taiwo, B.B.A. (1988) Production Characteristics of a Herd of West African Dwarf Goats at Ubiaja, Bendel State of Nigeria. In: Smith, O.B. and Bosman, H.G., Eds., Goat Production in the Humid Tropics, Proceedings of a Workshop at the University of Ife, Ile-Ife, 20-24 July 1987, 140-144.

28 Lawrence, T.L. and Fowler, V.R. (1997) Growth of Farm Animals. CAB International, Wallingford, 330. Cited by Atta, M. and El Khidr, O.A. (2004) Use of Heart Girth, Wither Height and Scapulo-Ischial Length or the Prediction of Live Weight of Nilotic Sheep. Small Ruminant Research, 55, 233-237.

29 Yakubu, A., Salako, A.E. and Imumorin, I.G. (2011) Comparative Multivariate Analysis of Biometric Traits of WAD and Red Sokoto Goats. Tropical Animal Health and Production, 43, 561-566.

30 Wilson, R.T., Muraiyi, T. and Rocha, A. (1989) Indigenous African Small Ruminant Strains with Potentially High Re- productive Performance. Small Ruminant Research, 2, 107-117.

31 Odubote, I.K. and Akinokun, J.O. (1992) Estimates of Genetic Parameters for Economic Traits in West African Dwarf Goats. Nigerian Journal of Animal Production, 19, 114-119.

32 Adebambo, O.A., Samuel, A.O. and Onakade, A.D. (1994) Causes of Variation in Reproductive Performance of West African Dwarf Goat. Nigerian Journal of Animal Production, 21, 29-36.

33 Odubote, I.K. (1996) Genetic Analysis of the Reproductive Performance of West African Dwarf Goats in the Humid Tropics. In:

Lebbie, S.H.B. and Kagwini, E., Eds., Small Ruminant Research & Development in Africa, Proceedings of the 3rd Biennial Congress of the African Small Ruminant Research Network, Kampala, 5-9 December 1994, 33-36.

34 Odubote, I.K. and Akinokun, J.O. (1994) Estimates of Phenotypic and Genetic Correlations in West African Dwarf Goats. Nigerian Journal of Animal Production, 21, 47-50.

35 Thilagam, K., Ramamoorthi, J., Sivaselvam, S.N., Karthickeyan, S.M.K. and Thangaraju, P. (2006) Kanniadu Goats of Tamilnadu, India: Genetic Characterization through Microsatellite Markers. Livestock Research for Rural Development, 18.http:// www.cipav.org.co/lrrd/lrrd18/10/thill18149.htm

36 Adebambo, A.O., Williams, J.L., Blott, S. and Urquart, B. (2011) Genetic Distance between Two Popular Nigerian Goat Breeds Used for Milk Production. Livestock Research for Rural Development, 23. http://www.lrrd.org/lrrd23/2/adeb23.26.htm

37 Yakubu, A., Salako, A.E., De Donato, M., Takeet, I., Peters, S.O., Adefenwa, M.A., Okpeku, M., Wheto, M., Agaviezor, B.O., Sanni, T.M., Ajayi, O.O., Onasanya, G.O., Ekundayo, O.J., Ilori, B.M., Amusan, S.A. and Imumorin, I.G. (2013) Genetic Diversity in Exon 2 of the Major Histocompatibility Complex Class II DQB1 Locus in Nigerian Goats. Biochemical Genetics, 51, 954-966.

38 Zarate, A.V. (1996) Breeding Strategies for Marginal Regions in the Tropics and Sub-Tropics. Animal Research & Development, 43-44, 100-118.

39 Kosgey, I.S. (2004) Breeding Objectives and Breeding Strategies for Small Ruminants in the Tropics. Ph.D. Thesis, Animal Breeding & Genetics Group, Wageningen University, Wageningen, 272 p.

40 Oseni, S.O., Ajayi, B.A. and Muritala, I. (2006) Mainstreaming Farmer Innovativeness in WAD Goat Genetic Improvement Schemes in the Development of Sustainable Goat Production Systems in South-Western Nigeria. In: Abstracts of Proceedings, Innovation Africa Symposium, Kampala, 20-23 November 2006, 75.

41 Musa, M.A., Peters, K.J. and Ahmed, M.K.A. (2006) On-Farm Charcaterization of Butana and Kenana Cattle Breeds and Production Systems in Sudan. Livestock Research for Rural Development, 18. http://www.lrrd.org/lrrd18/12/musa18177.htm

42 Kahi, A.K., Rewe, T.O. and Kosgey, I.S. (2005) Sustainable Community Based Organizations for the Genetic Improvement of Livestock in Developing Countries. Outlook on Agriculture, 34, 261-270.

43 Dossa, L.H., Wollny, C. and Gauly, M. (2007) Spatial Variation in Goat Populations from Benin as Revealed by Multivariate Analysis of Morphological Traits. Small Ruminant Research, 73, 150-159.

44 Reynolds, L. and Adediran, S. (1993) Composition of Village Goat Production Herds in Southwest Nigeria. Small Ruminant Research, 13, 49-53.

45 Terrisse, B.F. (2005) Improving the Performance of Indigenous Sheep Breeds. Leisa Magazine, September 2005, 16-17.

46 Oseni, S.O., Sonaiya, E.B., Omitogun, O.G., Ajayi, B.A. and Muritala, I. (2006) West African Dwarf Goat Production under Village Condition: 1. Characterization and the Establishment of Breed Standards. Proceedings of Conference on International Agricultural Research for Development, Book of Abstracts, Tropentag 2006, University of Bonn, Bonn, 11-13 October 2006, 526.

47 Tibbo, M. (2008) Animal Breeding in Developing Countries Context. NCCR Trade Regulation IP-9 Workshop on Animal Breeding, Innovation, Trade & Proprietary Rights, Berne, 27-28 November 2008, 45 p.

48 Yapi-Gnoare, C.V. (2000) The Open Nucleus Breeding Programme of the Djalloke sheep in Cote d'Ivoire. In: Galal, S., Boyazoglu, J. and Hammond, K., Eds., Workshop on Developing Breeding Strategies for Lower Input Animal Production Environments, ICAR Series 3, Bella, 22-25 September 1999, 283-292.

49 Yapi-Gnaore, C.V., Dagnogo, B. and Oya, B.A. (2003) Community-Based Livestock Improvement and Conservation: Experience from Open-Nucleus Breeding Programmes in West Africa. FAO, Rome, 89-95. http://www.fao.org/docrep/006/Y3970E/y3970e09.htm

# CHAPTER 6

# Some Peculiarities of Horse Breeding

**Marcilio Dias Silveira da Mota[1] and Luciana Correia de Almeida Regitano[2]**

[1] Department of Animal Breeding and Nutrition, School of Veterinary Medicine and Animal Science, University of Sao Paulo State, Botucatu/SP, Brazil

[2] Animal Molecular Genetics, Embrapa Southeast Cattle, Sao Carlos/SP, Brazil

## 1. INTRODUCTION

There are approximately 60 million horses in the world, most of them living in America, Asia and some European countries. Currently China has the largest herd (around 8 million), followed by the United States (7 million), Mexico (a little more than 6 million) and Brazil (a little less than 6 million). Together, these four countries have close to 45% of the world's equine population [1].

The horse population's growth rate has been either constant or decreasing in most of the countries, with only some regions in Central America,

Asia and Europe keeping positive growth rates (Figure 1). From 2003 to 2007, Puerto Rico presented the highest growth rate (45.4%), while the biggest Decrease was in Benin, Africa (-12.9%) (Figure 2).

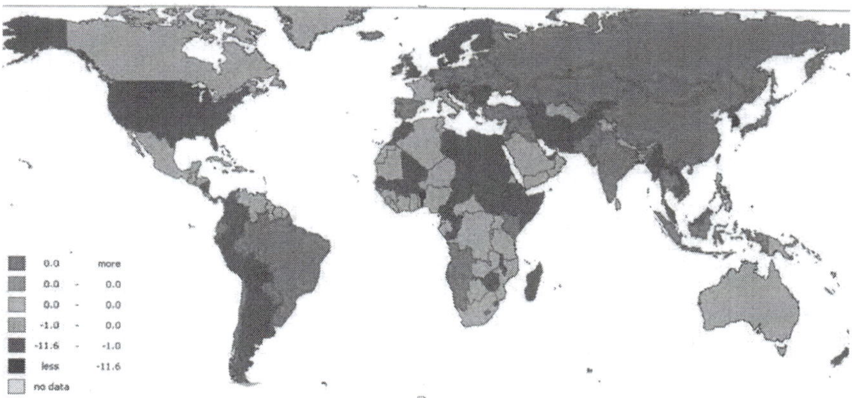


**Figure 1.** Growth rate of horse population in the world from 1997 to 2007 (Source [1])

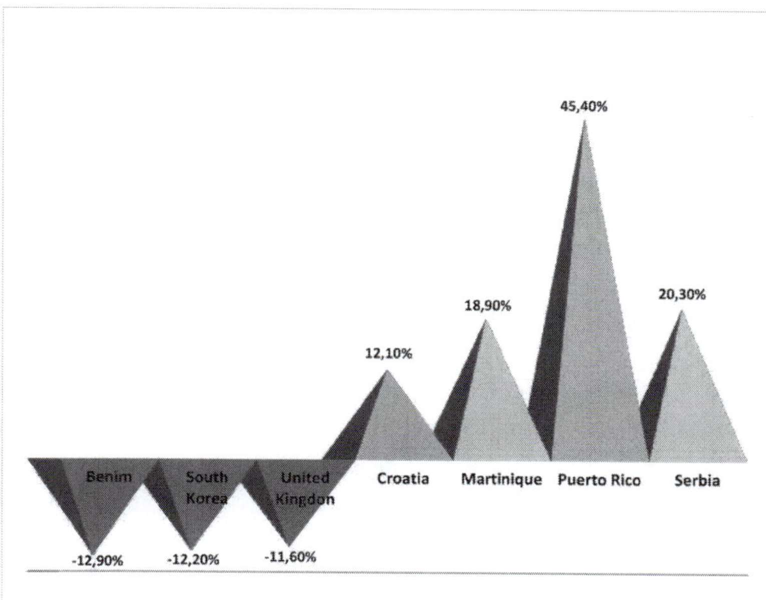


**Figure 2.** Countries with higher and lower equine population growth rates from 2003 to 2007

The world's export trade of live horses is concentrated in Europe and America, together representing 84% (58% and 26% respectively) of 2.1 billion dollars moved around yearly [2]. According to this source, the United States were the main exporter (148,472 animals, representing 48.8% of the world trade), raising around 474 million, in average $ 3,200 per animal.

Similar to the exportation, the world trade of live horses in relation to imports is highly supported by Europe (49.8% of the world trade), followed by Asia (26%) and Mexico (82,854), Canada (63,240) and Italy (46,333) are the main importers in number of animals, although the highest global expenditure in this aspect occurred in the United Kingdom (U.S. $ 498 million), UAE ($ 236 million) and Ireland (U.S. $ 233 million), representing together 46% of annual turnover - [2].

## 2. HORSE BREEDING

The expression "improvement of equine species," according to [3] dates back to the French articles about horses and donkeys in the*"Histoire Naturelle"* by Buffon in 1753. Despite the strong creationist dogmas of that time, these articles anticipated evolutionist ideas, while they described the concept of race degeneration due to the influence of location or climate. Adopted by the French Veterinary schools, it was believed that in order to return to an ideal type of creation, it would be necessary to mate mares and stallions with opposite types, to compensate traits that differed from the ideal. This meant basically to look for stallions on distant lands or regions.

Only decades later, after Darwin's evolution theory introduced the concept of selection is that the improvement of native populations emerged to counterbalance the situation, and effectively replace the desire to keep local types. Later, and gradually, the management of purebred animals predominated over crossbreeding, ending in the Thoroughbred Studbook's creation (first worldwide). It was a very important step for the use of animal production techniques in the 19th century [4]. According to the author, these techniques required precise identification procedures records of races held at different times and places in order to relate them to the same horse. The Stud Book, from the animal lineage certification, not only made it possible to relate the own horse performance information, but also their relatives. This fact prepared the way for the pedigree selection and progeny tests.

In addition, until the beginning of last century, horses were the focuses of experimental tests of inheritage theories. In this sense, historically the main concepts introduced by the horses, especially by the Thoroughbred breed, were [4]: performance selection in a purebred, introducing of a precise method for identifying an animal and its relatives, thinking about the male and female roles in the herd and the widespread use of planned matings.

Currently, although research in several countries are published every year in the literature involving some kinds of horse breeding study, few of them have ongoing consistent selection program. It means, in a way that most of research results in this area don't generate practical application and therefore it adds little to the species' development.

In most cases, this fact occurs because the breeders don't show interest in using the research results and not because of the research quality. The studies, depending on the availability and quality of information, usually do not consider the breeders' interests. Thus, in most countries there is wide gap between research institutes / universities and breeder associations. This gap is particularly due to the fact that more than in other species, horse breeders consider other breeders as potential competitors and give little importance to joint actions needed in breeding programs.

On the other hand, closer relation in Japan, Canada and some European countries, have allowed major advances in horse breeding. In Germany, for example, a country that stands out for jumping and dressage competitions and export of horses, the opinion, awareness and cooperation among breeders gave associations set guidelines that would meet not only their needs, and those of the research institutes / universities, but also the interest of the country. So the Breeders' Associations, supported by these institutions, publishes annually a Sire Handbook containing productive traits considered of general interest. This liaison has been very productive, resulting in19 gold, 8 silver and 12 bronze medals only in the last two Olympics[1] -, besides over 50 medals in some recent world championships [5].

# 3. SOME ADVANTAGES AND DIFFICULTIES IN HORSE BREEDING

Comparing to other farm animal species, there are some advantages and difficulties specifically related to researches on horse breeding, when quantitative genetics principles are used.

The advantages are fundamentally the amount of performance information and pedigree extension.

- The depth of the genealogy in most breeds is high

Traditionally, horse breeders associations consider the "pedigree" as a key factor to select their animals, so that in most of them, herd control is efficient. This fact goes back to the world's first Stud Book opening in 1791 (General Stud Book) for Thoroughbred animals. It was the basis for the other*Stud*Books, not only for horses, but also for all other domestic species.

Most of the breeds that were studied based on breeding, it is possible to track the animal's genealogy back to the fourth or fifth generation (depending on when the breeders' association's started), so that genetic evaluations in this aspect are efficient.

- Economic important traits can generally be measured in both genders.

In most of the breeds, especially the ones for some kinds of performance, the possibility of measuring both genders generates a greater information and knowledge volume about the behavior of traits in the population, what enables more efficient genetic evaluations. Racing, jumping, dressage, barrel, performances[2] - are examples of this kind.

On the other hand, even in breeds (or strains within a breed) where the breeding economic interest is the production of animals only for the conformation, not for performance, both genders can be evaluated.

- Economic important traits can usually be measured repeatedly in short time periods.

In horses it is possible to get repeated performance in relatively short time periods in a large portion of the population, while that doesn't happen with some domestic species, where economic important traits aren't repeated in the animal life (weight at weaning, weight at the year), or require a relatively long time to repeat (milk production, weight of the fleece).

In this sense, considering that the average number of starts per horse in Thoroughbred race season 2007/2008 was 6.5 [6], in a year is possible to get reasonable performance information (position, awards, time) about the horses. Although this average represents a new performance every two months, there are animals that race every 15 days, or even during the same week.

On the other hand, certain difficulties related to equine species and others due to various issues have been some of the major problems when research on horse breeding is done.

## 3.1. Inherent To The Species

- Low reproductive rates

Since the early ancestors emerged from 55 to 60 million years ago, horses are adapting in order to develop a reproduction model that ensures survivability in the wild, adopting different reproductive strategies to ensure that their progeny are born in the appropriate time of year [7]. However, domestication has strongly influenced reproductive performance, with selection pressure on fertility being either small or null, and mating usually dictated by the functional performance of the animals [8].

Thus, considering that reproductive traits usually have low heritability estimate [8], and have been selected on horses by an indirect way, genetic alterations in order to enhance characteristics of this nature are evidently neither fast nor simple, especially in a species with a long generation interval such as the equine's (see item "High ranges of generation and delivery").

Hence, horses have low reproductive performance when compared to other farm animal species. The birth rate ranges from 59% [9, 8] to 74% [10, 7] the higher percentages being usually found in tests involving a small number of mares. Table 1 illustrates the result of 42,750 matings done with 7,278 Thoroughbred mares. It was observed that birthrates for males and females were 49.26% and 50.74%, respectively, whereas abortion and stillborn foals were 1.41% and 2.02%, respectively. It was also recorded that 9.07% of coverings were classified as empty, whereas 23.7% of matings did not show any latter records to track success.

**Table 1.** Occurrences after mating

| Occurrence | Number of observations and % |
|---|---|
| Male | (29.35%) |
| Female | 12,927 (30.23%) |
| Abortion | 603 (1.41%) |
| Empty | 3,878 (9.07%) |
| Without information | 9,864 (7.23%) |
| Stillborn | 866 (2.02%) |
| Not mated | 1,945 ( 4.54%) |
| Mated with another breed | 135 (0.31%) |

Source: [8]

This relatively low fertility is described in different races, countries and purposes, and may be related to hormonal dysfunction, genital infections in mares, parasitic infestations and inadequate handling practices before the breeding season [11]. These factors are even more imposing in the case of animals used in sports, since they have very different handling from those that are exclusively used in breeding [12]. Furthermore, maintenance of older mares and stallions due to their progeny's superior sporting performance can decrease the rate of conception in the herd, as the rate decreases progressively with increasing age [13]. Table 2 describes this aspect, illustrating the conception and apparent fertility rates (defined by [14] as the ratio of the total number of mares that conceived by the total number of mated mares and as mares that delivered living foals by the number of mated mares, respectively) according to parturition order.

Additionally, in some breeds (Thoroughbred, Quarter Horse, Standardbred, etc.), the existence of the so-called Racing Year[3] - also contributes to the decrease in reproductive rates. This is because, in certain competitions, animals are usually grouped according to, among other parameters (awards, past performance, gender, etc.), the horse's age defined by the racing year. Thus, breeders try to get the products born closer to the beginning of the equestrian year (July 1$^{st}$ - southern hemisphere or January 1st - northern hemisphere) in order to seize a competitive advantage (better developed, mature and trained horses) in relation to the animals born later that year [15]. Figure 3 shows the concentration of births of Thoroughbred foals in Brazil, a country located in the southern hemisphere.

**Table 2.** Number of observations (N), conception rates and apparent fertility according to the parturition order of Thoroughbred mares.

| Parturition Order | N | Conception rate (%) | Apparent fertility rate (%) |
|---|---|---|---|
| 1 | 5.531 | 75 | 71 |
| 2 | 4.534 | 73 | 68 |
| 3 | 3.655 | 71 | 68 |
| 4 | 2.794 | 69 | 67 |
| 5 | 2.155 | 67 | 65 |
| 6 | 1.560 | 64 | 61 |
| 7 | 1.070 | 62 | 59 |
| 8 | 777 | 59 | 56 |
| 9 | 513 | 56 | 54 |
| 10 | 329 | 54 | 52 |
| 11 | 171 | 48 | 47 |
| 12 | 92 | 48 | 46 |
| > 13 | 79 | 41 | 38 |

Source: [8]

With that objective, the breeding season, that usually lasts for 4 to 5 months, usually starts in August 15th in the southern hemisphere (or February 15th in the northern hemisphere). However, the percentage of mares that naturally ovulate in this period is quite low, since mares are seasonal poliestric, with the onset of the natural breeding season in spring, associated with increases in photoperiod, temperature and food availability [7]

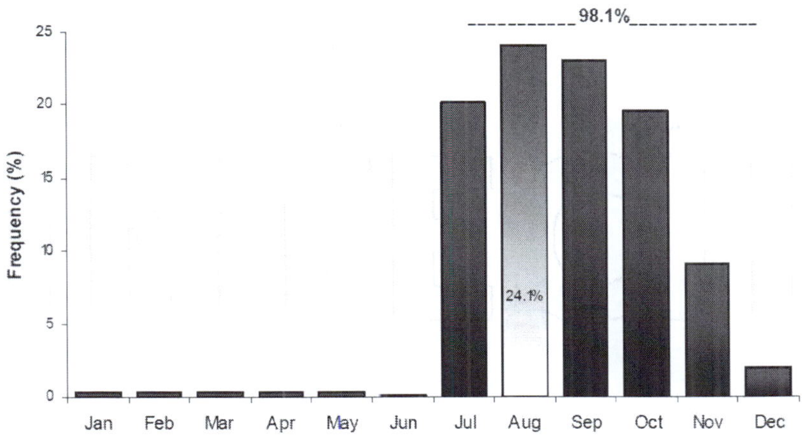


**Figure 3.** Percentage of Thoroughbred foal births by month in Brazil

In this sense, [15] found in Thoroughbred horses raised in Ireland (Northern Hemisphere), that the change of the beginning of the breeding season from February 15th to April 15th (hence delaying the beginning of the racing year by the same amount - January 1 to March 1), would better accommodate the natural reproductive cycle of females and could potentially increase the pregnancy rate by approximately 10% (Figure 4). Similar gains were likely to occur in countries in the southern hemisphere if the beginning of the breeding season changed from August 15th to October 15th, consecutively postponing the beginning of the racing year from July 1st to September 1st. Following this guidance, Australia postponed the beginning of the national racing year to August 1st (breeding season between September and December), mitigating the problem, although [13] considers the best breeding season in that country to be between November and February.

Figure 5 helps to understand the idea suggested by [13], as it represents the photoperiod in the southern hemisphere according to the latitude. It is observed that photoperiods are longer between October and February and higher percentages of ovulating mares are expected, favoring successful coverings.

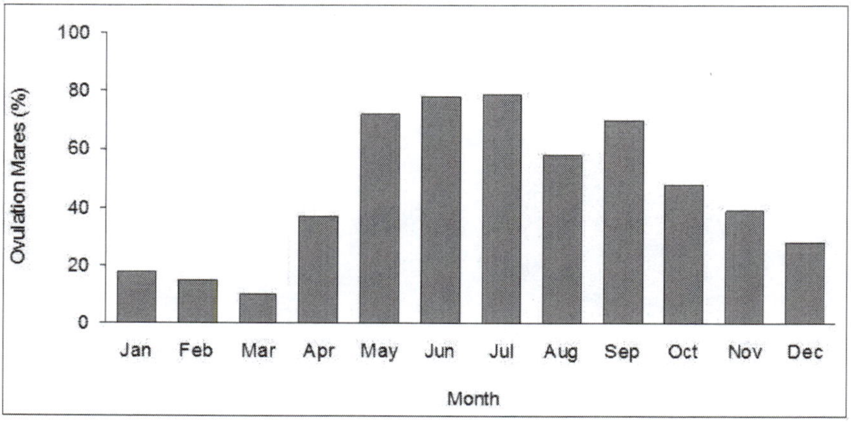


**Figure 4.** Percentage of mares ovulating by month in the northern hemisphere (adapted from [15])

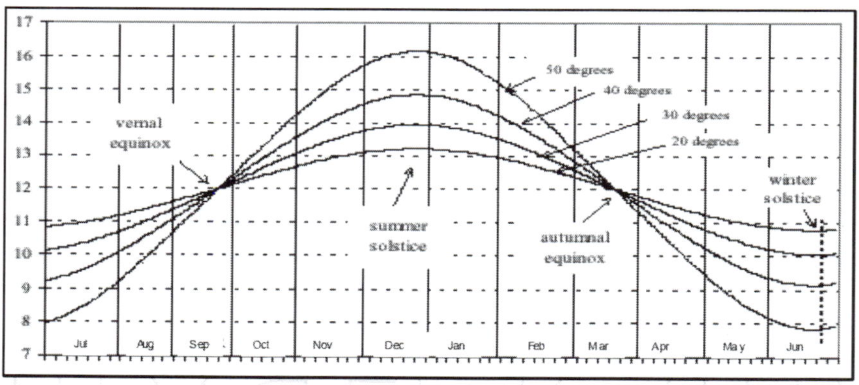


**Figure 5.** Photoperiods for different degrees south latitude (adapted from [16])

- high generation and parturition intervals

Generation interval represents the time needed to replace the next generation, and the shorter it is, the greater the expected annual gene change rate. Thus, considering that in horses this interval varies from 8 to 12 years, overall genetic changes due to selection tend to be slower when compared to cattle (4-6 years), sheep (3-5 years) pigs (1.5 to 2 years) and birds (1 to 1.5 years). Sportive breeds, in which reproductive technologies are not permitted, typically show higher ranges as the superior performance animals, especially females, usually start into reproduction after the end of their competitive life.

The generation intervals of some equine breeds are described in Table 3.

**Table 3.** Generation intervals from a few equine breeds

| Breed | Generation Interval (years) | Reference |
|---|---|---|
| Andalusian | 10.1 | [17] |
| Thoroughbred | 7.10 | [18] |
| Icelandic Toeler | 9.7 | [14] |
| Friesian | 9.6 | [19] |
| Hanoverian | 8.4 | [20] |
| Mangalarga | 9.5 | [21] |
| Arabian | 9.7 | [22] |
| Lusitano | 10.5 | [23] |
| Mangalarga Marchador | 8.9 | [24] |

The parturition interval is the amount of time between two consecutive parturitions, including the time from the parturition until the appearance of the first heat, from the first heat to the conception and finally the duration of pregnancy. It is an important component when estimating the herds' reproductive efficiency, with great influence on the economic return and breeding, due to its effects on the generation interval and selection intensity to be applied.

Among the domestic species, the mare has the capacity to provide fertile estrus a few days after birth, the so called foal heat. The main advantage of this phenomenon seems to be the maintenance of a 12 months foaling interval [25]. According to these authors, due to the enrollment of horses in sport activities, there is great pressure in order to have as many pregnant mares as possible in the breeding season. In this sense, efforts are made to cover mares in the foal heat. Considering the average pregnancy lasting around 11 months (see topics below) and the possibility of new pregnancy in the close postpartum days, a 12 months foaling interval would be obtained [25]. Thus, most breeders seek to take the opportunity of the foal heat, being aware that during this heat mares ovulate quickly, conception rates are lower and early embryonic mortality rate is higher [8].

There is a great variation among the equine foaling intervals, depending on the breed and breeding purposes, although most studies point values greater than 365 days (Table 4).

**Table 4.** Foaling intervals in some equine breeds

| Breed | Foaling interval (days) | Reference |
|---|---|---|
| Halflinger | 468 | [26] |
| Thoroughbred | 490 | [8] |
| Marwari | 535 | [27] |
| Kathiawari | 567 | [27] |
| Mangalarga Marchador | 548 | [24] |
| Arabian | 387 | [7] |

- Relatively long gestation period and low number of offspring per parturition

Although the duration of a pregnancy isn't directly associated with a breeding farm's production costs, its study may be extremely important in the preparation of breeding plans.

The gestation period can be defined as the time between fertilization of egg and the fetus' delivery. According to [28], the average duration of a mare's pregnancy is typically 340 days, ranging from 300 to 400 days. This wide time range until the birth of the foals indicates that mares may be highly susceptible to both internal and external factors afecting the duration of the pregnancy [29]. The ages of mare and stallion, year and month of birth, breeding season, foal sex, breed and nutritional status are factors that should be considered in the study of the pregnancy's duration [28].

Studies focusing the stallion used on coverings also deserve special attention when studying the pregnancy period in mares. The pregnancy's duration for females mated with specific stallions may be a criterion when choosing the stallion. This is because when a mare is bred late in a breeding season, yet the owner wants to mate her during this season, choosing a stallion associated with shorter pregnancy durations may be profitable [28].

Working with horses in the northern hemisphere [30] observed that the mating season was the most important factor affecting the duration of pregnancy in mares. According to these authors, the pregnancies that derived from mating during the period from December to May were 10.4 days longer than those derived from mating from June to November.

Studying Arab mares in Egypt, [31] observed that pregnancies with longer durations were the ones that ended in the winter, suggesting that the mares seem to be able to adapt the length of the gestation so the births happen in spring, which may be important for the survival of the species in the wild.

**Table 5.** Gestation periods in some equine breeds

| Breed | Gestation period (days) | Reference |
|---|---|---|
| Carthusian | 332 | [32] |
| Thoroughbred/Quarter Horse | 341 | [33] |
| Andalusian | 336 | [34] |
| Standardbred | 349 | [35] |
| Mangalarga Marchador | 327 | [24] |
| Arabian | 334 | [7] |
| Criollo | 335 | [36] |
| Freiberger | 336 | [37] |

Regardless of the discrepancy between studies and breeds, mares have a relatively long gestation period compared to other domestic species such as cattle (270-290 days), goats (145-151 days), sheep (144-152 days), pigs (112-115 days) and buffalo (298-317 days). Furthermore, as a uniparous species, twins (or multiple) are rare in mares, incidence varying from 0.5 to 1.6% of parturitions [38], so that the annual availability of animals for selection is comparatively small.

## 3.2. Other Aspects

- Use of reproductive technologies

Although reproductive technologies (artificial insemination and embryo transfer) make possible different practical advantages such as lower disease and injury transmission, long-term storage of genetic material, easier transportation, earlier onset of reproduction in females and, in the case of embryo transfer, reproduction during the sports career, in the context of animal breeding they are usually considered additional tools to optimize breeding programs [5].

The commercial impact of these techniques in horses varies greatly. For example, in Thoroughbred horses meant for racing, the use of artificial insemination[4] - "The fecundation of mares can only be made by direct sexual contact, not admitting artificial insemination, but may exceptionally be authorized by the Brazilian Stud Book, by virtue of proven physical impairment of the player, the use of immediate reinforcement with fresh semen collected during the coverage. This procedure, when authorized, will be held only by a veterinarian authorized in advance by the Brazilian Stud Book " is officially banned worldwide, while in several sportive breeds, especially in Europe, the percentage of inseminated animals exceeds 80% (French Saddle Horse - 84%, Hanoverian - 91% Holsteiner - 95%, Belgian and Dutch Warmblood - over 95%) [5].

On the other hand, according to the International Embryo Transfer Society (IETS) the higher number of embryo transfers occur in the United States, Argentina and Brazil, in order of importance, countries which together represent over 90% of the activities in this area worldwide. In 2005, 5,700 embryos were transferred in Brazil, none frozen, while across Europe only 711 transfers were done in horses [39].

In this context, it is observed that, although several countries present a prominent role in the worldwide scenario with respect to the use of these reproductive biotechniques, especially embryo transfer, there are few studies about their impact (generation interval, accuracy, selective intensity, inbreeding) on horses breeding programs. Research in this subject would be important to understand the direction that has been given to these tools in different countries and monitor their actual benefit for the horse population.

- Knowledge of interesting traits' economic values

A fundamental requirement for any breeding scheme aiming the improvement of quantitative traits is the establishment of the breeding objectives, involving the relative values of genetic change for all desirable features included in the breeding program. Typically, these values are expressed in monetary terms as weights to be applied to each feature of economic importance [40].

However, few scientific studies have been performed in order to obtain economic weights for traits involved in horse selection programs, were a combination of empirical experience, some biological factors and intuition of designers prevail. There are a few reasons for that.

It's often very difficult to determine, in horses, the value of one unit of expression for a given trait in relation to the animal's total value. Economically quantifying units for traits such as speed, dressage, jumping, etc., is far more complex than attributing values to liters of milk, kilos of meat or wool.

The long period of time between mating and expressig the traits of interest in the progeny, besides the difficulties in determining an appropriate function for profitability, provide part of this deficiency. Moreover, according to recent authors, another problem arises from the fact that not always the relative economic weights are linear in a breeding program. Thus, the amount of increase in the genotype for certain character can be strongly dependent on the values of other genetic traits. For example, in horses with outstanding ability to jump, the additional genetic values affecting their training capability is almost neglected, whereas in animals with a low ability to jump, a genotype corresponding to training characteristics can greatly increase its value.

- Diversity of goals within certain breeds

There are equine breeds that are commonly selected by breeder in only one direction, as is the case of the Thoroughbred, where the objective is basically to obtain animals with superior performance in races. In Quarter horses, in addition to races, performance in work tests and conformation may also be targets of breeders since the race is subdivided amongst these strains.

Moreover, in breeds in which animals are involved in a wide variety of uses (work on farms, horseback riding, trekking, exhibitions, equine therapy, equestrian tourism, unskilled riding sports, etc.) breeders seek very different traits, depending on the purpose of raising the animal, hindering the implementation of breeding programs that cover all segments. Brazilian breeds such as the Mangalarga and the Mangalarga Marchador fall into this category.

In these cases, studies involving quantitative and molecular aspects of traits that can meet the desire for the greater proportion of breeders can result in meaningful contributions to selection of reproductive, behavioral, immunological and other traits.

# NOTES

[1] - taking into account dressage disciplines, jumping and eventing.

[2] - meets three classical disciplines: dressage, cross country and jumping. It is a form held in three days.

[3] - In the southern hemisphere, an interval of 12 months between July 1st and June 31st. In northern hemisphere countries, begins on January 1st and ends in December 31st.

[4] - Indeed, in exceptional cases may occur insemination as set forth in Article 25 of the Rules of the Association of Breeders and Owners Horse Racing - Stud Book.

# REFERENCES

1. GHILPA2011The Global Livestock Production and Health Atlas. Available: http://kids.fao.org/glipha/.Acessed 2011 November 20.
2. FAO2011Food and Agriculture Organization of the United Nations.

Available: http://faostat.fao.org/site/569/default.aspx#ancor.Acessed 2011 November 22.

3. J. Mulliez, 1983Les chevaux du royaume. Ed Montalba, 385p.

4. B. Langlois, 1996A consideration of the genetic aspects of some current practices in Thoroughbred horse breeding. Ann. Zootech. 454151

5. Bruns EW2008Horse Breeding in Germany. In: Horse breeding and genetics course, class notes. Graduation Program on Animal Breeding and Genetics, Unesp/FCAV, Jaboticabal, 45 p.

6. IFHA2012International Federation of Horseracing Authorities, Statistical Information, Available: www.IFHAOnline.org.Acessed 2012 May 5.

7. S. Cilek, 2009The Survey of reproductive success in Arabian horse breeding from 1976-2007 at Anadolu State Farm in Turkey. Journal of Animal and Veterinary Advances 8389396

8. R. Z. Taveira, M. D. S. Mota, 2007Genetic and quantitative evaluation of breeding traits in Thoroughbred mares. Revista Eletrónica de Veterinaria 8111

9. Mota MDS2005Avaliação populacional da raça Mangalarga. In: Reunião do Conselho Técnico da Associação Brasileira de Criadores do Cavalo Mangalarga, Ribeirão Preto 63p.

10. Zúccari CESN, Nunes DB, Costa e Silva EV2004Harém Pantaneiro sob monta natural a campo na Região do Rio Negro, Pantanal, MS. In: Anais, IV Simpósio sobre Recursos Naturais e Sócio-econômicos do Pantanal. Corumbá, 14

11. J. J. Sullivan, P. C. Turner, L. C. Self, H. B. Gutteridge, D. E. Bartlett, 1975Survey of reproductive efficiency in the Quarter-Horse and Thoroughbred. J. Reprod. Fert. Suppl. 23315318

12. Jackson RS1971Equine infertility. Preliminary report on a survey of courses taken by the A.A.E.P. Newsletter Am. Ass. Equine Pract.

13. Bailey CJ1998Wastage in the Australian Thoroughbred Racing Industry. Rural Industries Research & Development Corporation, 98p.

14. K. Hugason, T. H. Arnason, J. V. Jónmundsson, 1985A note on the fertility and some demographical parameters of Icelandic toelter horses. Livest. Prod. Sci. 12161167

15. Cunningham EP1991The Genetics of Thoroughbred Horses. Scientific American 2649298

16. D. Diven, 2000Keys to the low cost/calf program. Missouri forage and grassland council Annual meeting, Lake Ozark, Missouri, 2000. Available:http://agebb.missouri.edu/mfgc/2000mtg/lowcost.htm. Acessed 2011 November 22.

17. M. Valera, A. Molina, J. P. Gutiérrez, J. Gómez, F. Goyache, 2005Pedigree analysis in the Andalusian horse: Population structure, genetic variability and influence of the Carthusian strain. Livestock Production Science: 955766

18. Taveira RZ, Mota MDS, Oliveira HN2004Population parameter in Brazilian Thoroughbred. J. Anim. Breed. Genet. 121384391

19. M. Sevinga, T. Vrijenhoek, J. W. Hesselink, H. W. Barkema, A. F. Groen, 2004Effect of inbreeding on the incidence of retained placenta in Friesian horses. Journal of Animal Science, 82982986

20. Hamann. H. [20], O. Distl, 2008Genetic variability in Hanoverian Warmblood horses using pedigree analysis. Journal of Animal Science 86150313

21. M. D. S. Mota, Prado. R. S. Almeida, J. Sobreiro, 2006 Caracterização da população de cavalos Mangalarga no Brasil. Archivos de Zootecnia, 553137

22. S. Moureaux, E. Verrier, A. Ricard, J. C. Meriaux, 1996Genetic variability within French race and riding horse breeds from genealogical data and blood marker polymorphism. Genetics Selection and Evolution 2883102

23. Carolino. N. Vicente, L. T. Gama, 2012Genetic diversity in the Lusitano horse breed assessed by pedigree analysis. Livestock Science- in press.

24. Gonçalves RW, Costa MD, Rocha Júnior VR, Costa MR, Silva ESP, Ribeiro AMF2011Inbreeding effect on reproductive traits in a herd of Mangalarga Marchador Brazilian horses. Rev. Bras. Saúde Prod. Anim., Salvador 12641649

25. Lowis TC, Hyland JH1991Analysis of post-partum fertility in mares on a Thoroughbred stud in southern Victoria. Australian Veterinary Journal 68, n.9, September.

26. A. Meregalli, C. Valzania, 1984Reproductive parameters of Halflingers kept on Tuscan farms. Rivista di Zootecnia e Veterinaria 1219098

27. Singh MK, Yadav MP, Mehta NT2002Breed characteristics of Marwari and Kathiawari horses. Indian Journal of Animal Sciences 72319323

28. Marteniuk JV, Carleton CL, Lloyd JW, Shea ME1998Association of sex of fetus, sire, month of conception, or year of foaling with duration of gestation in Standarbred mares. Javma 212: 11

29. Davies Morel MCG, Newcombe JR, HollandSJ(2002Factors affecting gestation length in the Thoroughbred mare. Animal Reproduction Science 74175185

30. Rollins WC, Howell CE1951Genetic sources of variation in the gestation length of the horse. Journal Animal Science 10: 797.

31. V. Hura, J. Hajurka, J. Kacmárik, G. Csicsai, I. Valocky, 1997The effect of some factors on gestation length in Nonius breed mares in Slovakia. UVL, 04101 Kosice, Komenského 73, Slovak Republic.

32. K. Satué, M. Felipe, J. Mota, A. Munoz, 2011Gestational length in Carthusian broodmares:effects of breeding season, foal gender, age of mare, year of parturition, parity and sire. Polish Journal of Veterinary Sciences 14173180

33. McCue PM, Ferris RA2012Parturition, dystocia and foal survival: A retrospective study of 1047 births. Equine Veterinary Journal 44, Suppl. 412225

34. M. Valera, F. Blesa, Santos. R. Dos, A. Molina, 2006Genetic study of gestation length in Andalusian and Arabian mares. Animal Reproduction Science 957596

35. M. Dicken, E. K. Gee, C. W. Rogers, I. G. Mayhew, 2012Gestation length and occurrence of daytime foaling of Standardbred mares on two stud farms in New Zealand. New Zealand Veterinary Journal 604246

36. Winter GHZ, Rubin MIB, De La Corte FD, Silva CAM2007Gestational Length and First Postpartum Ovulation of Criollo Mares on a Stud Farm in Southern Brazil. Journal of Equine Veterinary Science 27531534

37. R. Giger, H. P. Meier, U. Küpfer, 1997Gestation lengths of Freiberger» mares with mule and horse foals. Schweizer Archiv fur Tierheilkunde, 139303307

38. Basrur PK1983Genetics in Veterinary Medicine. University of Guelph, Guelph, Canada, 261p.

39. M. Thibier, 2006Transfers of both in vivo derived and in vitro produced embryos in cattle still on the rise and contrasted trends in other species in 2005. International Embryo Transfer Society Newsletter, 241218Available: http://www.iets.org/pdf/data_retrieval/december2006.pdf. Acessed: 2010 August 2.

40. T. Arnason, L. D. Van Vleck, 2000Genetic Improvement of the Horse. In: The Genetics of the horse. Cab Publising. 473497

# CHAPTER 7

# Genomic selection needs to be carefully assessed to meet specific requirements in livestock breeding programs

**Elisabeth Jonas*** **and Dirk-Jan de Koning**

Department of Animal Breeding and Genetics,
Swedish University of Agricultural Sciences, Uppsala, Sweden

## ABSTRACT

Genomic selection is a promising development in agriculture, aiming improved production by exploiting molecular genetic markers to design novel breeding programs and to develop new markers-based models for genetic evaluation. It opens opportunities for research, as novel algorithms and lab methodologies are developed. Genomic selection can be applied in many breeds and species. Further research on the implementation of genomic selection (GS) in breeding programs is highly desirable not only for the common good, but also the private sector (breeding companies). It has been projected that this approach will improve selection routines, especially in species with long reproduction cycles, late or sex-limited or expensive trait recording and for complex traits. The task of integrating GS into existing breeding programs is, however, not straightforward. Despite successful integration into breeding programs for dairy cattle, it has yet to be shown how much emphasis can be given to the genomic information and how much additional phenotypic information is needed from new selection candidates. Genomic selection is already part of future planning in many breeding companies of pigs and beef cattle among others, but further research is needed to fully estimate how effective the use of genomic information will be for the prediction of the performance of future breeding stock. Genomic prediction of production in crossbreeding and across-breed schemes, costs and choice of individuals for genotyping are reasons for a reluctance to fully rely on genomic information for selection decisions. Breeding objectives are highly dependent on the industry and the additional gain when using genomic information has to be considered carefully. This review synthesizes some of the suggested approaches in selected livestock species including cattle, pig, chicken, and fish. It outlines tasks to help understanding possible consequences when applying genomic information in breeding scenarios.breeding, estimated breeding value, marker-assisted selection, generation interval, modeling, non-additive effects

# 1. INTRODUCTION

Plant and livestock breeding started many 1000 years ago with the cultivation of plants and capture of individual animals, leading to the domestication of most agricultural species used today. Early approaches were based on phenotypic evaluations with little awareness of the underlying causes of different productivities and appearances. This was changed substantially when more information on the genetic background became available, firstly without and later by using genetic marker information. Basic methods were developed after the (re-)discovery of the Mendelian laws of inheritance, and further advances were based on the knowledge on inheritance of genetic information via linkage and linkage disequilibrium. Theoretical approaches and progression on the field of quantitative genetics, especially the work from Fisher, Wright and Lush, allowed multifactorial models and inclusion of complex pedigrees into defined breeding decisions (for example reviewed in Hill and Mackay, 2004; Hill, 2014). A combination of theoretical approaches and experimental achievements has played a significant role for the development of modern breeds. Today, structured breeding programs exist for most livestock species. They rely on the routine recording of pedigree and performance information on populations and they further implement the basic knowledge on inheritance into selection choices for the progression of the selected population. Breeding programs for different species and production systems apply different strategies and are built in different ways. Many livestock breeding populations (e.g., chicken, fish, pig, beef cattle) are structured in tiers of nucleus, multiplier and commercial farms, which are commonly shaped as a pyramid with few nucleus breeders and many commercial breeders (Nicholas, 2009). Genetic progress does mainly occur in the upper tier, and improved breeding stock is transferred by relatively few nucleus breeders to the multipliers. The flow of genes is directed downward and commercial farmers rely on both improvement in the nucleus herds as well as little loss of genetic progress during the multiplication of animals (Nicholas, 2009). Open-nucleus schemes have the advantage of a lower inbreeding rate and potentially increased selection response compared to a closed-nucleus. In a closed nucleus, the breeding company maintains the full ownership of the elite breeding material.

Livestock breeding programs use estimated breeding values (EBVs), mainly estimated for sires, for selection decisions based on own performance and that of relatives. EBVs are estimated using traits measured in an own-performance, pedigree, sib, and/or progeny evaluation scheme in specialized testing stations and/or selected farm environments using, most of the times, best linear unbiased predictions (BLUP; Henderson, 1975). The breeder's equation is an indicator for the change due to selection ($\Delta Z = h^2 \times$ S, where $\Delta Z$ is the change of the mean of a quantitative trait in one generation, heritability $h^2$ describing the phenotypic similarity between relatives or trait variation due to additive genetic effects, and S measures the selection intensity). It indicates that the change of a quantitative trait depends on its heritability and selection intensity (Falconer and Mackay, 1996; Lynch and Walsh, 1998; Xu and Hu, 2010). The evaluation of the genetic merit targets the sires in many livestock species (i.e., in cattle or pigs). Sires have more offspring, especially when artificial insemination (AI) is used, and their genetic impact on livestock populations is therefore stronger (Gerrits et al., 2005; Funk, 2006). AI stations and/or nucleus breeders play therefore the main role for improvement of the genetic merit in a population. Generation intervals of 4 to 5 years in beef cattle, more than 5 years in dairy cattle (Figure 1) and 2 to 2.5 years in pig breeding schemes are one consequence of the time-consuming evaluation of phenotypes to enable a reliable estimation of breeding values (Schefers and Weigel, 2012). The advances of having sophisticated breeding programs implemented, are diminished by the long time period until a new generation is available for breeding (often defined as the average age of selected parents when the offspring is born or generation interval), and costs for trait recording.

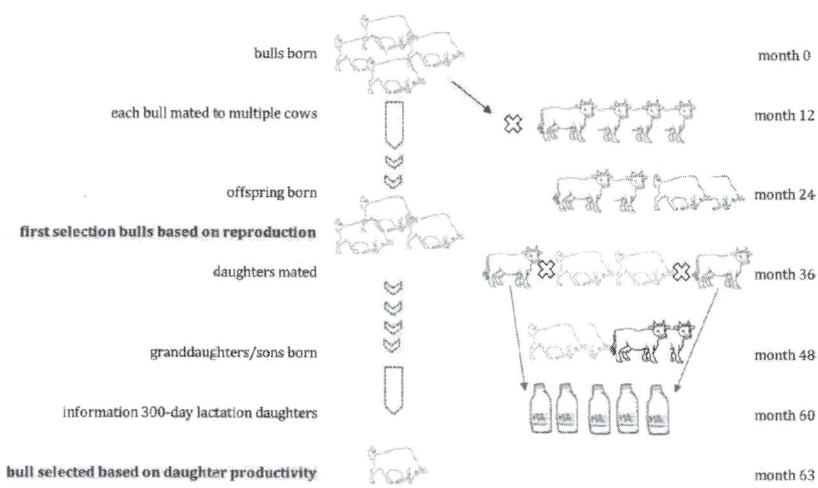


**FIGURE 1** Schematic overview of traditional breeding in dairy cattle: Selection candidates (bulls) are born and mated to multiple cows at around 12 months of age. In month 60 the first phenotypic recordings (milk yields) are available from the daughters of the selection candidate, as they had to reach puberty, be mated and finished a first lactation. Estimated breeding values are available after approximately 63 months and the bulls for further breeding are selected.

## 2. BREEDING USING MOLECULAR GENETIC MARKERS

The goal of most breeding programs is to predict the genetic merit of an individual and thus allow targeted combinations of desired alleles to improve the performance of the next generation(s). The phenotype of an individual is only in rare cases a good indicator for allelic differences, while the use of genetic markers allows tracing detailed information on the inherited part of the genome, other than such observed by the phenotype (Botstein et al., 1980). Today, genetic DNA markers are used to assist breeding and selection and to build the basis for novel breeding approaches such as marker-assisted selection (MAS) and more recently GS (Williams, 2005). MAS, which uses one or few markers as a selection tool, has not reached the initial proposed achievements, the main reason is that the discovery of reliable markers is difficult, especially when working with complex traits. Only relatively few consistent markers could be identified, despite successful identification of many quantitative trait loci (QTL; Hu and Reecy, 2007). Advancements of molecular genetic tools, allowing genotyping of samples at many loci at a time, further advanced the aim of using genetic markers for selection. Meuwissen et al. (2001) suggested GS, which is based on the use of genome-wide markers panels to derive breeding values based on genomic information. Genomic EBVs (GEBVs) or predictions of the genetic merit of an individual based on its genome are derived for candidates with genotype and phenotype information, the so-called training population. The information is then used for the selection of genotyped candidates with no recorded traits (selection candidates). GS can be applied in practice for all main livestock species since genome-wide single nucleotide polymorphism (SNP) panels or even full sequence information are available (Table 1). GS has been incredibly successful in dairy cattle, where GEBVs are published in a number of countries[1]. Different prediction methods (linear models such as BayesA, BayesB, GBLUP etc., non-linear models such as Neural Networks or Reproducing Kernel Hilbert spaces; penalized methods including ridge regression or Bayesian shrinkage estimation) applied to a wide range of datasets have been reviewed previously and suggestions of implementations were discussed and can be found elsewhere (e.g., Daetwyler et al., 2012; de Los Campos et al., 2012; Morota and

Gianola, 2014). However, beside general prediction methods also some hurdles to be taken before this selection approach can be of practical use in livestock populations other than dairy cattle have been described (Ibanez-Escriche and Gonzalez-Recio, 2011). Especially the basic breeding programs, accuracies of EBVs and possibilities and needs of implementing non-additive effects into genomic predictions differ between populations. This might also have an effect on marker density required and thus genotyping costs, additionally might the value of each individual selection candidate be relatively low, compared to an elite breeding bull (Ibanez-Escriche and Gonzalez-Recio, 2011).

**Table 1** Overview of genome structure and genotyping platforms of the main livestock species.

| Species | Ploidy | Genome size [Mb] | Number of genes | Genome structure | Linkage disequilibrium | Commercial arrays | Reference sequence |
|---|---|---|---|---|---|---|---|
| Cattle (*Bos taurus*) | Diploid (2n = 60) | ~2870 | ~26835 | 3000000 SNP identified | Highly variable extent of LD | Illumina: 54609 SNP, Affymetrix: 640000 SNP | Hereford cow (whole genome shotgun), her sire (hierarchical BAC clone; Elsik et al., 2009) |
| Pig (*Sus scrofa*) | Diploid (2n = 38) | ~2596 | ~21640 | 510000 SNP identified | Higher LD (than some Holstein cattle) | Illumina: 64232 SNP | Female domestic Duroc pig (Illumina whole-genome shotgun, BAC clone; Groenen et al., 2012) |
| Chicken (*Gallus gallus*) | Diploid (2n = 78) | ~1000 | 20000–23000 | 1800000 SNP identified | Difference of LD between layer lines | Affymetrix: 580000 SNP | Single red jungle fowl female from inbred line; ~6.6 × whole-genome, shotgun reads BAC-end read pairs (Int Chicken Genome Sequencing, 2004) |
| Atlantic salmon (*Salmo salar*) | Diploid (2n = 58) | ~6000 | 33709 (identified in 2010) | Many chromosomal rearrangements | Moderate LD | iSelect Atlantic salmon 16,500 SNP | Female fish; aimed end of 2013 (Davidson et al., 2010); announced finish June 2014 |

*SNP, single nucleotide polymorphism; LD, linkage disequilibrium; BAC, bacterial artificial chromosome.*

# 3. APPLICATION OF GENOMIC SELECTION IN DAIRY CATTLE

Dairy cattle are mostly bred based on a rather simple within-breed selection principle. Occasionally crossbred calves are produced which can then be used for beef production if no replacement heifers are needed. Selection is mainly performed on the sire-side; bulls have a much greater contribution to the national and international breeding stock as their semen is distributed via AI. Despite the success of breeding programs in improving traits such as milk yield, limitations of traditional selection have been suggested. Increased health and fertility problems, the relative importance of dairy cattle production world-wide, the long generation interval due to the long reproductive cycle and the indirect measurement of the main traits of interest via the daughters of a bull made dairy cattle a good model for the implementation of GS. Further early efforts to design genotyping panels and to assemble the full genome sequence (Table 1), as well as the relatively closely linked world-wide population of the main breed, Holstein-Friesian, were reasons that GS was firstly employed in dairy cattle. An additional benefit was that GS could be implemented within the existing breeding and recording structures. Costs and generation intervals of traditional breeding using progeny testing are considerable and the improvements provided when using GS were prospected to have an incredible impact on the dairy industry in the future (Hayes et al., 2009; Thomasen, 2013).

Generation intervals were suggested to drop from around 5 to 6 years in traditional dairy cattle breeding programs to around 1.5 years (Figure 2) when using GS (Pryce and Daetwyler, 2012). It has been predicted that the costs for testing of bulls can be drastically reduced as GEBVs are available at birth, leading to an improved genetic change among other possible advances (Schaeffer, 2006). Costs for genotyping and thus the implementation of GS in dairy cattle is also cheaper since mainly male animals are being genotyped and semen of bulls is often distributed across multiple countries. Reasons for that are that AI is commonly used in dairy cattle populations, semen storage over a long time is possible in cryopreservation, and preferred bulls have therefore a higher impact on the breeding stock. Many breeding programs focus on the selection of sires while using phenotypic information from cows only for the derivation of EBVs (and GEBVs) for bulls. Only a few breeding programs include cows into their reference population to better assess additional traits and increase the reference population (Figure 3).

Genomic selection has been introduced into dairy cattle breeding programs rapidly and GEBVs are published in a number of countries (see text footnote 1). The integration of GS into breeding schemes has been reviewed recently (Pryce and Daetwyler, 2012; Bouquet and Juga, 2013). There are, in general, two options for the integration of genome-wide selection into breeding decisions, by either pre-selecting young bulls for further testing (pre-selection scheme) or for selection of new breeding bulls based on genomic information only (turbo scheme). However, especially due to the reduced generation interval (as in the "turbo" scheme), genomic inbreeding rates need to be observed and managed. The pre-selection scheme offers a rather conservative method with no change to the breeding program while producing moderate genetic gain, while the turbo scheme allows high genetic gain by considerably changing the structure of the breeding program (Bouquet and Juga, 2013). Details of the design of commercial breeding schemes with implemented GS are mostly unknown as most breeding companies might keep and test their strategy first in detail as this methodology has yet to be fully established (Pryce and Daetwyler, 2012).

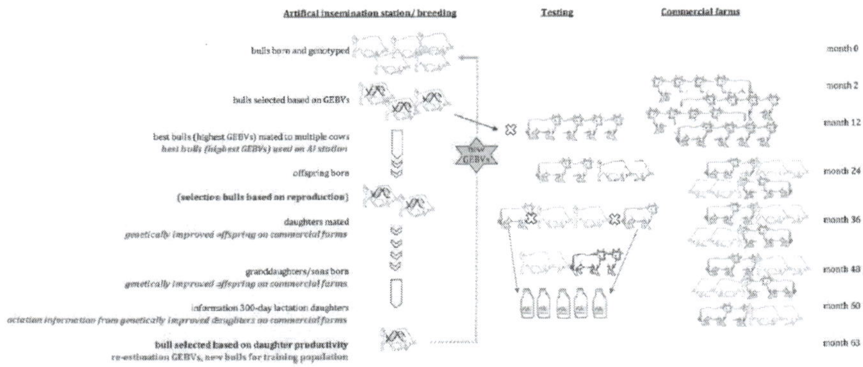


**Figure 2** Marker enhanced breeding using Genomic Selection in dairy cattle: Genome-wide marker panels are used to identify association or linkage with traits of interest using bulls with information on daughter lactation and genotypic data. Estimated effects of each marker are summarized into genomic estimated breeding values (GEBV), which can be used for selection.The figure assumes that GEBVs are available from an existing training population in month 0 and are being updated in month 63 with additional information. Only selected animals (based on GEBVs) are used for testing. Semen of superior bulls can be used earlier on commercial farms, allowing for high ergenetic gain.Genomics election allows a significant reduction of the generation interval from 6 to 1.5years if applie das suggested in the figure.

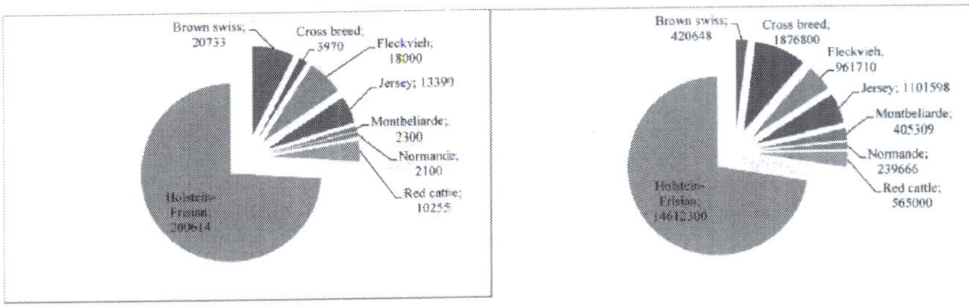


**Figure 3** Numbers of animals in active genomic selection breeding programs adapted from (Thomasen, 2013). Shown is the total number of bulls in the reference panel **(A)** and recorded cows **(B)**.

# 4. APPROACHES IN BEEF CATTLE

In beef cattle breeding, selection indices are often based on a specified market. Economic weights given to each selection target assist to combine the EBVs for different traits for the selection of the breeding stock. EBVs are, depending on the trait and number of information used, relatively good predictors for the future productivity of breeding stock. However, current breeding values are still less accurate compared to those in dairy cattle (Johnston et al., 2012). Genetic markers are commercially licensed for some traits in beef cattle, but many markers used for trait improvement do not perform consistently across populations (Allan and Smith, 2008). The use of genome-wide marker sets has therefore been of major interest for beef cattle breeding. Traditional breeding using progeny testing focusses largely on bulls at the nucleus level. An extended inclusion of cow information into the selection process would allow farmers to select superior dams (Saatchi et al., 2012) and lead to a breeding program which allows a more balanced inclusion of all traits considered in the breeding goal (Garrick, 2011). However using GS to select stock, especially cows, on the farm level is often restricted to purebred operations and pre-selection of cows would still be required to keep costs at a reasonable level. Only a cost-efficient strategy, for example by providing high accuracy of prediction, will allow successful use of GS also in beef cattle populations (Johnston et al., 2012). Heritability of the economically most important traits, training population (or total number of bulls with reliable EBVs) as well as the effective population size are factors influencing the applicability of genetic markers in beef cattle and determine also the great differences to the currently observed successes in dairy cattle (Johnston et al., 2012).

Overcoming especially the problem of small training populations would require a combination of data across countries and/or across breeds, and thus, higher density marker panels would be required to allow reliable predictions (de Roos et al., 2009). It has been hypothesized that a higher marker density, which would allow the inclusion of causative mutations or markers perfectly inherited with such, would improve predictions in less related populations (Pollak et al., 2012). However, relevance of marker density, size and relationship within and across training and selection population are still being discussed. Additionally, beef cattle population are less uniform compared to dairy cattle and both *Bos taurus* and *Bos indicus* are being used (Garrick, 2011). Better collation of genotype data, cross-comparison with large training populations (using higher density marker panels), possibly inclusion of females into the training data set and more emphasis on phenotyping might improve the potential for GS in beef cattle.

## 5. IMPLEMENTATION OF GENOMIC SELECTION IN PIGS

In pigs, performance tests for offspring from selection candidates are completed in testing stations to reduce interactions through environmental factors. Animals are, in some countries such as Germany or Sweden, additionally tested on commercial farms to integrate variable environments into the selection process. Improvement of the genetic merit of pig populations is mainly based on sire selection and AI is commonly used. Most breeding schemes use crossbreeding between paternal production-oriented breeds and maternal reproduction-oriented breeds, and selection is mainly performed in purebred nucleus herds (Visscher et al., 2000). Sires lines are selected for improved carcass and meat traits, aiming to predict the performance of crossbred offspring.

Genomic selection in pig breeding is a potential tool to improve maternal traits as traditional breeding programs focus on performance traits in the sire lines. A simulation study has shown that GS has the potential to increase genetic gain for maternal traits in pigs (Lillehammer et al., 2011). A simulation study for the implementation of GS in male lines also concluded that the phenotypic selection can be improved while being (based on the selected genotyping strategy) economically efficient (Tribout et al., 2012).

Genomic selection might also assist to increase genetic gain in developing countries in which indigenous and exotic lines could be used for repeated backcrossing (Akanno et al., 2013). The use of genomic information will, also in such schemes, assist to control inbreeding and avoid a possible loss of variation.

Lower genetic correlations between traits in cross- and purebred animals (Dekkers, 2007) are reasons that GS has not yet been used as a selection tool with the same reliability as in dairy cattle. Many production traits are measured in crossbred pigs, while selection is based on the purebred lines they derived from. The integration of QTL information into breeding decision might enable to predict non-additive effects. However, inclusion of data from crossbred animals is required to reach higher accuracy compared to phenotypic selection (Dekkers and Chakraborty, 2004). Biologically meaningful markers, such as causative genes, candidate genes within gene-pathways or markers in strong linkage disequilibrium (LD) with the causative mutation might overcome the poor prediction of crossbred performance using information from purebreds only. It has been suggested that the use of MAS might assist to increase accuracies and thus genetic gain as it overcomes the difficulties especially for traits which are genetically different in purebred and commercial crossbred pig populations, for example those influenced by the environment (Dekkers, 2007).

Economic motivations are relevant for a successful implementation of GS in pig breeding, the reduction of the generation interval is significantly less compared to dairy cattle and current traditional selection methods do work well, despite the high costs for progeny testing. In 2014, one large, world-wide running pig breeding company announced their use of GS in crossbred populations[2]. However no details of the prediction method and accuracies have been given.

# 6. SUGGESTED BREEDING PROGRAMS IN CHICKEN

Chicken breeding programs are in a pyramid form and comparable to pig breeding (Dekkers, 2007). The larger number of offspring in chicken allows more than double the genetic improvement compared to cattle or pigs when using traditional breeding. The implementation of GS in chicken has been discussed as it would reduce the generation interval from 12 to 6 months. It has been stated that costs will be a major part of the decision making process in scenarios for which genotypes are added to the existing performance testing scheme. Changes to most of the current breeding programs have been suggested, before a cost-efficient implementation of GS can be realized. The main change is the reduction of the number of animals within the program while keeping the effective population size the same (Sitzenstock et al., 2013). *In silico* studies have also suggested the implementation of GS in broiler lines, for which genotyping strategies need to be chosen carefully to reduce costs but still provide the full information (Avendaño et al., 2010). Also a layer breeding program with 250 males and females (compared to 1000 males and 3000 females in traditional breeding) would still lead to reasonably good predictions (Avendaño et al., 2010). Using a similar scenario, with 293 males and 913 females across four generations in the training population did also illustrate the potential of GS (Wolc et al., 2010).

But theoretical studies could, until now, not provide a clear answer on how genomic breeding values will be predicted in cross-breeding schemes, which are also commonly used in chicken breeding. It is possible that predictions made in genetically relatively close purebred lines will give reliable estimates for improvements in crossbreds, if QTL are inherited with the same marker allelic variants. However, another explanation for differences between predictions from purebred in crossbred populations is the difference of environmental factors. In one of the scenarios suggested by Sitzenstock et al. (2013) crossbred hens were therefore chosen from the environment closest to the production system, since housing conditions in purebred lines are often highly standardized.

However, side-effects could be a problem in most scenarios tested so far: one study observed a decreased egg weight, possibly as the consequence of improved feed efficiency of laying hens (Sitzenstock et al., 2013). Those scenarios also suggest that breeding organizations need to decide if faster genetic improvement of the breeding population will be a strong enough argument for a genomic enhanced, but more costly, selection scheme (Sitzenstock et al., 2013), taking possible risk of unwanted effects. It is in general important to carefully define breeding goals and indices to minimize negative effects on other economically relevant traits.

# 7. CONSTRAINS IN FISH

Aquaculture is a broad field and combining hundreds of species under one header might not reflect the real potential and importance of aquaculture for provision of food. Also the opportunities and breeding programs vary widely in aquaculture as do generation intervals. There are only a few reports on GS in aquaculture in general and among the most important farmed fish species, studies are currently only published for the Atlantic salmon (Sonesson and Meuwissen, 2009; Nielsen et al., 2011; Lillehammer et al., 2013). The transfer of genetic gain from nucleus breeder to multiplier to commercial farms is, due to the high reproduction capability in fish, high and fast. The structure of the Norwegian Atlantic salmon breeding scheme allows approximately 8–10% genetic gain per generation for some traits (Gjoen and Bentsen, 1997). However, inbreeding might be higher if fewer breeding animals are kept (Gjedrem, 2005).

Current family-based breeding schemes in aquaculture, using information from close relatives (e.g., sibs) to estimate breeding values include only approximately half of the genetic variation into the selection decision (Sonesson and Meuwissen, 2009). A simulation study testing a GS-based sib-testing breeding scheme generally reduced the total genetic gain. It was suggested that aquaculture breeding programs would need to be re-designed to accommodate GS as the high genotyping costs could only be covered if very high genetic gain can be achieved (Sonesson and Meuwissen, 2009). Such changes could for example include fewer families, reduced phenotypic evaluation or use of field data from commercial farms, and either higher costs or reduced genetic gain have to be considered as consequences (Sonesson and Meuwissen, 2009; Nielsen et al., 2011). Achieved genetic gain would be higher at increased costs due to genotyping compared to conventional BLUP based breeding (Sonesson and Meuwissen, 2009). Breeding programs could be optimized in a way that pre-selection based on easy measureable but economic relevant traits like growth, would reduce the number of selection candidates (Sonesson and Meuwissen, 2009). It has been shown that GS controls inbreeding more effectively while allowing for improved genetic gain in a sib-based compared to a traditional BLUP breeding scheme (Nielsen et al., 2011). Reasons for increased inbreeding in BLUP breeding schemes are the co-selection of sibs in sib designs due to truncation selection on EBVs (Sonesson and Meuwissen, 2009), greater reproductive capacity of males and females, and inability to use within-family variation (Nielsen et al., 2011).

A combination between traditional BLUP estimation, pre-selection of candidates and low-density genotyping arrays might be one possibility to overcome the difficulties of high costs without high loss of genetic gain (Lillehammer et al., 2013). Alternative strategies, such as phenotypic recording from double haploids, which are applicable to fish populations, have been simulated and suggested as additional test population to gain genetic information. It has been shown that a careful design of test versus selection population for example using double haploids has the potential for increased selection accuracy compared to traditional sib testing schemes especially for traits that are difficult to record such as disease resistance or filet quality (Nirea et al., 2012).

# 8. SUMMARY AND CONCLUSION

We reviewed here the approaches of GS in different livestock species and identify restrictions based on current breeding schemes. This article might provide a basis for the critical reading of articles on GS and thus interpretation of reported results and accuracies. This in turn could contribute to a broader range of viewpoints in future articles. There are more complex aspects to the topic of GS and issues mentioned here, such as assignment of varying effect sizes to markers, modeling of gene-by-gene or genotype-by-environment interactions or use of imputed genotypes and/or phenotypes. Further detailed analysis will be needed to allow a comprehensive demonstration on how to solve them in breeding schemes and selection decisions.

The title of this review emphasizes the diversity of current breeding schemes, and the need to either implement GS into existing structures or change breeding plans to fit novel breeding tools as suggested by others in more detail (e.g., Henryon et al., 2014; Ibanez-Escriche et al., 2014; Van Eenennaam et al., 2014). We summarized that even in dairy cattle populations, for which GS is widely applied, different schemes for the implementation of GS are being offered, allowing a faster or more conservative selection strategy. One of the main issue is that genomic predictions need to be reliable over many generations and the long-term efficiency of GS has yet to be proven in most practical applications. Tools for GS are available and genetic marker information can be implemented into breeding programs, combined with information on phenotypes and pedigrees of larger groups of individuals via complex algorithms. However, more questions have arisen with increased knowledge of the theoretic background and options of methodologies and technologies.

Some breeding practices make use of non-additive effects such as heterosis in cross-breeding. Genetic markers are often restricted to the prediction of additive effects, more complex mechanisms which cannot be explained by the DNA sequence information will be ignored when using such information for a breeding decision. Discussion will continue on the relevance of marker density, population size and structure to allow reliable estimations of the QTL effects and possibly diminish the size of not-explained effects by using genetic markers. Other uncertainties in the prediction of individual productivity or phenotypic appearance can occur due to inbreeding depression. And finally, the combination of (non-homozygote) genetic information is generated at random and combinations of non-linked loci can only be predicted in terms of probabilities.

The extensive number of publications and studies on models and methods for GS in different species indicates the potential of this still relatively novel breeding tool. Nevertheless, further research collaborations have arisen from some of the challenges that were identified, including the need for denser marker sets, which can now be further augmented since full genome sequences of many bulls are being collated (Daetwyler et al., 2014). Further studies using more applied populations and relevant traits need to proof if the expectations can be fulfilled. Despite some of the concerns, such as possibly higher costs due to genotyping not met by economic gains, GS has a future, since it holds more advantages such as control of inbreeding and known heritable defects or functional mutations, which again do add secondary advantages for marketing and product branding. A linked review on GS in crop species in the same research topic, aims to further deepen the relevance for an extended community.

## AUTHOR CONTRIBUTIONS

EJ and D-JK discussed the different aspects of genomic Selection; EJ wrote the first draft of the paper, while the final manuscript was written in collaboration.

## CONFLICT OF INTEREST STATEMENT

The authors declare that the research was conducted in the absence of any commercial or financial relationships that could be construed as a potential conflict of interest.

## ACKNOWLEDGMENTS

This work was financially supported by Mistra Biotech, a research program financed by Mistra – Stiftelsen för miljöstrategisk forskning and SLU.

# FOOTNOTES

[1]http://www.interbull.org [accessed January 31, 2013]

[2]http://www.pic.com/cms/USA/796.html

# REFERENCES

1 Akanno E. C., Schenkel F. S., Sargolzaei M., Friendship R. M., Robinson J. A. B. (2013).Opportunities for genome-wide selection for pig breeding in developing countries. J. Anim. Sci. 91, 4617–4627. 10.2527/jas.2013-6102

2 Allan M. F., Smith T. P. L. (2008). Present and future applications of DNA technologies to improve beef production. Meat Sci. 80, 79–85. 10.1016/j.meatsci.2008.05.023

3 Avendaño S., Watson K. A., Kranis A. (2010). "Genomics in poultry breeding—from utopias to deliverables," in Proceedings of the 9th World Congress on Genetics Applied to Livestock Production (WCGALP), Leipzig.

4 Botstein D., White R. L., Skolnick M., Davis R. W. (1980). Construction of a genetic linkage map in man using restriction fragment length polymorphisms. Am. J. Hum. Genet. 32, 314–331.

5 Bouquet A., Juga J. (2013). Integrating genomic selection into dairy cattle breeding programmes: a review. Animal 7, 705–713. 10.1017/s1751731112002248

6 Daetwyler H. D., Calus M. P. L., Pong-Wong R., de los Campos G., Hickey J. M. (2012). Genomic prediction in animals and plants: simulation of data, validation, reporting and benchmarking. Genetics93, 347–365 10.1534/genetics.112.147983

7 Daetwyler H. D., Capitan A., Pausch H., Stothard P., Van Binsbergen R., Brondum R. F., et al. (2014). Whole-genome

sequencing of 234 bulls facilitates mapping of monogenic and complex traits in cattle. Nat. Genet. 46, 858–865. 10.1038/ng.3034

8 Davidson W., Koop B., Jones S., Iturra P., Vidal R., Maass A., et al. (2010). Sequencing the genome of the Atlantic salmon (*Salmo salar*). Genome Biol. 11, 403. 10.1186/gb-2010-11-9-403

9 Dekkers J. C. M. (2007). Marker-assisted selection for commercial crossbred performance. J. Anim. Sci. 85, 2104–2114. 10.2527/jas.2006-683

10 Dekkers J. C. M., Chakraborty R. (2004). Optimizing purebred selection for crossbred performance using QTL with different degrees of dominance. Genet. Sel. Evol. 36, 297–324. 10.1186/1297-9686-36-3-297

11 de los Campos G., Hickey J. M., Pong-Wong R., Daetwyler H. D., Calus M. P. L. (2012). Whole genome regression and prediction methods applied to plant and animal breeding. Genetics 193, 327–345. 10.1534/genetics.112.143313

12 de Roos A. P. W., Hayes B. J., Goddard M. E. (2009). Reliability of genomic predictions across multiple populations. Genetics 183, 1545–1553. 10.1534/genetics.109.104935

13 Elsik C. G., Tellam R. L., Worley K. C., Gibbs R. A., Muzny D. M., Weinstock G. M., et al. (2009).The genome sequence of taurine cattle: a window to ruminant biology and evolution. Science 324, 522–528. 10.1126/science.1169588

14 Falconer D., Mackay T. (1996). Quantitative Genetics. Harrow: Longman.

15 Funk D. A. (2006). Major advances in globalization and consolidation of the artificial insemination industry. J. Dairy Sci. 89, 1362–1368 10.3168/jds.S0022-0302(06)72203-2

16 Garrick D. J. (2011). The nature, scope and impact of genomic prediction in beef cattle in the United States. Genet. Sel. Evol. 43, 17. 10.1186/1297-9686-43-17

17 Gerrits R. J., Lunney J. K., Johnson L. A., Pursel V. G., Kraeling R. R., Rohrer G. A., et al. (2005).Perspectives for artificial insemination and genomics to improve global swine populations.Theriogenology 63, 283–299. 10.1016/j.theriogenology.2004.09.013

18 Gjedrem T. (2005). Selection and Breeding Programs in Aquaculture. Berlin: Springer.

19 Gjoen H. M., Bentsen H. B. (1997). Past, present, and future of genetic improvement in salmon aquaculture. ICES J. Mar. Sci. 54, 1009–1014 10.1016/S1054-3139(97)80005-7

20 Groenen M. A. M., Archibald A. L., Uenishi H., Tuggle C. K., Takeuchi Y., Rothschild M. F., et al. (2012). Analyses of pig genomes provide insight into porcine demography and evolution. Nature 491, 393–398. 10.1038/nature11622

21 Hayes B., Bowman P., Chamberlain A., Goddard M. (2009). Invited review: genomic selection in dairy cattle: progress and challenges. J. Dairy Sci. 92, 433–443. 10.3168/jds.2008-1646

22 Henderson C. R. (1975). Best linear unbiased estimation and prediction under a selection model.Biometrics 31, 423–447. 10.2307/2529430

23 Henryon M., Berg P., Sørensen A. C. (2014). Animal-breeding schemes using genomic information need breeding plans designed to maximise long-term genetic gains. Livest. Sci. 166, 38–47 10.1016/j.livsci.2014.06.016

24 Hill W. G. (2014). Applications of population genetics to animal breeding, from Wright, Fisher and Lush to genomic prediction. Genetics 196, 1–16. 10.1534/genetics.112.147850

25 Hill W. G., Mackay T. F. C. (2004). D. S. Falconer and Introduction to quantitative genetics. Genetics167, 1529–1536.

26 Hu Z.-L., Reecy J. M. (2007). Animal QTLdb: beyond a repository. Mamm. Genome 18, 1–4. 10.1007/s00335-006-0105-8

27 Ibanez-Escriche N., Forni S., Noguera J. L., Varona L. (2014). Genomic information in pig breeding: science meets industry needs. Livest. Sci. 166, 94–100 10.1016/j.livsci.2014.05.020

28 Ibanez-Escriche N., Gonzalez-Recio O. (2011). Review. Promises, pitfalls and challenges of genomic selection in breeding programs. Span. J. Agric. Res. 9, 404–413 10.5424/sjar/20110902-447-10

29 Int Chicken Genome Sequencing C. (2004). Sequence and comparative analysis of the chicken genome provide unique perspectives on vertebrate evolution. Nature (Lond.) 432, 695–716. 10.1038/nature03154

30 Johnston D. J., Tier B., Graser H. U. (2012). Beef cattle breeding in Australia with genomics: opportunities and needs. Anim. Prod. Sci. 52, 100–106 10.1071/an11116

31 Lillehammer M., Meuwissen T. H. E., Sonesson A. K. (2011). Genomic selection for maternal traits in pigs. J. Anim. Sci. 89, 3908–3916. 10.2527/jas.2011-4044

32 Lillehammer M., Meuwissen T. H. E., Sonesson A. K. (2013). A low-marker density implementation of genomic selection in aquaculture using within-family genomic breeding values. Genet. Sel. Evol. 45 10.1186/1297-9686-45-39

33 Lynch M., Walsh B. (1998). Genetics and Analysis of Quantitative Traits. Sunderland, MA: Sinauer.

34 Meuwissen T. H. E., Hayes B. J., Goddard M. E. (2001). Prediction of total genetic value using genome-wide dense marker maps. Genetics 157, 1819–1829.

35 Morota G., Gianola D. (2014). Kernel-based whole-genome prediction of complex traits: a review.Front. Genet. 5:363. 10.3389/fgene.2014.00363

36 Nicholas F. W. (2009). Introduction to Veterinary Genetics. (West Sussex: Wiley-Blackwell; ), 328.

37 Nielsen H. M., Sonesson A. K., Meuwissen T. H. E. (2011). Optimum contribution selection using traditional best linear unbiased prediction and genomic breeding values in aquaculture breeding schemes. J. Anim. Sci. 89, 630–638. 10.2527/jas.2009-2731

38 Nirea K., Sonesson A., Woolliams J., Meuwissen T. (2012). Strategies for implementing genomic selection in family-based aquaculture breeding schemes: double haploid sib test populations. Genet. Sel. Evol. 44, 30. 10.1186/1297-9686-44-30

39 Pollak E. J., Bennett G. L., Snelling W. M., Thallman R. M., Kuehn L. A. (2012). Genomics and the global beef cattle industry. Anim. Prod. Sci. 52, 92–99 10.1071/an11120

40 Pryce J. E., Daetwyler H. D. (2012). Designing dairy cattle breeding schemes under genomic selection: a review of international research. Anim. Prod. Sci. 52, 107–114 10.1071/an11098

41 Saatchi M., Schnabel R. D., Rolf M. M., Taylor J. F., Garrick D. J. (2012). Accuracy of direct genomic breeding values for nationally evaluated traits in US Limousin and Simmental beef cattle.Genet. Sel. Evol. 44, 38. 10.1186/1297-9686-44-38

42 Schaeffer L. (2006). Strategy for applying genome-wide selection in dairy cattle. J. Anim. Breed. Genet. 123, 218–223. 10.1111/j.1439-0388.2006.00595.x

43 Schefers J. M., Weigel K. A. (2012). Genomic selection in dairy cattle: integration of DNA testing into breeding programs. Anim. Front. 2, 4–9 10.2527/af.2011-0032

44 Sitzenstock F., Ytournel F., Sharifi A., Cavero D., Taubert H., Preisinger R., et al. (2013). Efficiency of genomic selection in an established commercial layer breeding program. Genet. Sel. Evol. 45, 29. 10.1186/1297-9686-45-29

45 Sonesson A. K., Meuwissen T. H. E. (2009). Testing strategies for genomic selection in aquaculture breeding programs. Genet. Sel. Evol. 41, 37. 10.1186/1297-9686-41-37

46 Thomasen J. R. (2013). Genomic Selection in Small Dairy Cattle Populations. Ph.D. thesis, Aarhus University, Aarhus.

47 Tribout T., Larzul C., Phocas F. (2012). Efficiency of genomic selection in a purebred pig male line. J. Anim. Sci. 90, 4164–4176. 10.2527/jas.2012-5107

48 Van Eenennaam A. L., Weigel K. A., Young A. E., Cleveland M. A., Dekkers J. C. M. (2014).Applied animal genomics: results from the field. Annu. Rev. Anim. Biosci. 2, 105–139. 10.1146/annurev-animal-022513-114119

49 Visscher P., Pong-Wong R., Whittemore C., Haley C. (2000). Impact of biotechnology on (cross)breeding programmes in pigs. Livest. Prod. Sci. 65, 57–70 10.1016/S0301-6226(99)00180-3

50 Williams J. L. (2005). The use of marker-assisted selection in animal breeding and biotechnology.Rev. Sci. Tech. 24, 379–391.

51 Wolc A., Stricker C., Arango J., Settar P., Fulton J. E., O'Sullivan N., et al. (2010). "Breeding value prediction for production traits in layers using pedigree and marker based methods," in Proceedings of the 9th World Congress on Genetics Applied to Livestock Production (WCGALP), Leipzig.

52 Xu S., Hu Z. (2010). Methods of plant breeding in genome era. Genet. Res. 92, 423–441. 10.1017/s0016672310000583

# CHAPTER 8

# Breeding value prediction for production traits in layer chickens using pedigree or genomic relationships in a reduced animal model

**Anna Wolc,[1,2] Chris Stricker,[3] Jesus Arango,[4] Petek Settar,[4] Janet E Fulton,[4] Neil P O'Sullivan,[4] Rudolf Preisinger,[5] David Habier,[2] Rohan Fernando,[2] Dorian J Garrick,[2] Susan J Lamont,[2] and Jack CM Dekkers[2]**

1Department of Genetics and Animal Breeding, University of Life Sciences in Poznan, Wołyńska st. 33, 60-637 Poznan, Poland.

2Department of Animal Science, Iowa State University, Ames, IA 50011-3150, USA.

3Applied Genetics Network, Börtjstrasse 8b, 7260 Davos, Switzerland.

4Hy-Line International, Dallas Center, IA 50063, USA.

5Lohmann Tierzucht GmbH, 27472 Cuxhaven, Germany.

## ABSTRACT

### Background

Genomic selection involves breeding value estimation of selection candidates based on high-density SNP genotypes. To quantify the potential benefit of genomic selection, accuracies of estimated breeding values (EBV) obtained with different methods using pedigree or high-density SNP genotypes were evaluated and compared in a commercial layer chicken breeding line.

## Methods

The following traits were analyzed: egg production, egg weight, egg color, shell strength, age at sexual maturity, body weight, albumen height, and yolk weight. Predictions appropriate for early or late selection were compared. A total of 2,708 birds were genotyped for 23,356 segregating SNP, including 1,563 females with records. Phenotypes on relatives without genotypes were incorporated in the analysis (in total 13,049 production records).

The data were analyzed with a Reduced Animal Model using a relationship matrix based on pedigree data or on marker genotypes and with a Bayesian method using model averaging. Using a validation set that consisted of individuals from the generation following training, these methods were compared by correlating EBV with phenotypes corrected for fixed effects, selecting the top 30 individuals based on EBV and evaluating their mean phenotype, and by regressing phenotypes on EBV.

## Results

Using high-density SNP genotypes increased accuracies of EBV up to two-fold for selection at an early age and by up to 88% for selection at a later age. Accuracy increases at an early age can be mostly attributed to improved estimates of parental EBV for shell quality and egg production, while for other egg quality traits it is mostly due to improved estimates of Mendelian sampling effects. A relatively small number of markers was sufficient to explain most of the genetic variation for egg weight and body weight.

# 1. BACKGROUND

During the first decade of the 21st century, there has been a rapid development of genomic selection tools. Through the application of genomic selection [1], marker information from high-density SNP genotyping can increase prediction accuracies at a young age, shorten generation intervals and improve control of inbreeding [2], which should lead to higher genetic gain per year. Many simulation studies have shown the benefits of this technology, depending on heritability, number and distribution of effects of QTL, population structure, size of training data set used to estimate SNP effects, and other factors [3]. However, studies on real data are still scarce. If practical application of genomic selection is to be implemented in chicken breeding, as already done for dairy cattle [4], it must prove its advantage over traditional methods and be used in a way that maximizes the use of available information. The accuracy of EBV derived from large numbers of markers for within-breed selection is difficult to evaluate analytically and must be validated by correlating predictions to phenotype in the target population (usually the generation following training).

One of the challenges in genomic prediction of breeding values is that not all phenotyped individuals are genotyped. One approach to exploit all available information is to first estimate breeding values of genotyped individuals by pedigree-based methods using all data, including phenotypes on non-genotyped relatives, and then use deregressed estimates of those EBV for marker-based analyses [5,6]. This two-step approach may, however, result in suboptimal use of information. Another recently developed method uses a combined pedigree and genomic covariance matrix, which can incorporate both genotyped and non-genotyped animals [7,8]. However, these methods are computationally demanding and require careful scaling of the genomic relationship matrix to be consistent with the pedigree-based relationship matrix.

The reduced animal model was proposed by Quaas and Pollak [9] to make breeding value prediction under the animal model less computationally demanding. It fits the full relationship matrix for parents and absorbs the equations for non-parents. Nowadays, the development of powerful computers makes the reduction of computing cost less relevant for pedigree-based analyses but the reduced model can also be used to exploit marker-based relationships. In breeding programs using marker information, individuals that have been used for breeding (i.e. parents) are more likely to be genotyped than unselected non-parents. Estimating breeding values for genotyped animals and absorbing non-genotyped progeny into their equations can make full use of all available data. With this approach, there is no need to construct the inverse of the combined pedigree and genomic covariance matrix of Legarra et al. [7].

The objectives of this study were to implement a reduced animal model to estimate breeding values using high-density SNP genotypes, to evaluate the accuracy of breeding values estimated using high-density SNP genotypes in the generation following training in a layer breeding line, and to compare the accuracy of alternative methods of breeding value estimation.

# 2. METHODS

## 2.1 Data

Data on nine traits collected during the first 22 weeks of production were recorded on 13,049 birds from five consecutive generations in a single brown-egg layer line: egg production (ePD, percent hen average); age at sexual maturity (eSM, d); weight of the first three eggs laid by the hen (eE3, g) and shell color (eC3) collected from same eggs by Chroma Meter that measures lightness (L) and hue (as a function of a red-green (a) and a yellow-blue (b) scale). A second set of egg quality traits collected at 26-28 weeks (early, e) included average weight of eggs (eEW,g); egg color (eCO) eggs; shell quality measured as puncture score - a non-invasive deformation test averaged over points of the shell (ePS, Newton); albumen height (eAH, mm); and yolk weight (eYW, g). For birds selected on the basis of early (e) trait data, also late (l) production (42-46 weeks of age) traits were recorded: body weight (lBW, g); egg production (lPD, percent hen average); puncture score (lPS, Newton); egg weight (lEW, g); albumen height (lAH, mm); egg color (lCO, Lab); and yolk weight (lYW, g). Early and late egg quality measurements were averages of records on three to five eggs. In total 2,708 animals were genotyped for 23,356 segregating SNP (minor allele frequency >0.025;

maximum proportion of missing genotypes <0.05; maximum mismatch rate between parent-offspring pairs <0.05; parentage probability >0.95), using a custom high-density Illumina SNP panel. Of the genotyped animals, 1,563 were females with individual phenotypes and 1,145 were males without phenotypes. The genotyped set included sires and dams used for breeding in generations 1 to 5 and some progeny from generation 5. Breeding values were estimated for two stages of selection. To represent selection at a very young age, when own performances and phenotypes on female sibs were not yet available, training used all phenotypic data excluding generation 5, and validation was performed on 290 genotyped female individuals from generation 5. To represent selection of males at a later age, when phenotypes on female sibs are available, phenotypes of 2,167 non-genotyped hens from generation 5 were added to the training data but validation individuals were unchanged. A basic description of these data is given in Table 1.

## 2.2 Statistical analysis

Because of the data structure, a reduced animal model was applied with all parents genotyped and many non-genotyped non-parent progeny with phenotypes. In this approach, a distinction is made between genotyped individuals, including all parents, for which the full relationship matrix is fitted, and non-genotyped non-parent individuals. The following model was applied, following White et al. [10]:

$$y = Xb + (P + \frac{1}{2}QS + \frac{1}{2}QD)a + e$$

where

**y** is the ($N$x1) vector of observations,

**b** is the (25 × 1) vector of generation-hatch-line fixed effects,

**X** is the ($N$x25) incidence matrix for fixed effects,

**Table 1** Description of the population in terms of the number, mean and standard deviation of phenotypes by trait and generation

| Generation | | ePD | eEW | ePS | eAH | eCO | eE3 | eC3 | eYW | eSM | IBW | IPD | IEW | IPS | IAH | ICO | IYW |
|---|---|---|---|---|---|---|---|---|---|---|---|---|---|---|---|---|---|
| | N | 2,738 | 2,737 | 2,738 | 2,737 | 2,738 | 2,729 | 2,729 | 2,728 | 2,738 | 647 | 635 | 649 | 649 | 649 | 649 | 646 |
| G1Training | Mean | 80.93 | 56.81 | 1425 | 7.06 | 73.33 | 43.64 | 74.56 | 15.19 | 149.30 | 1.96 | 77.25 | 61.46 | 1,435 | 6.56 | 72.38 | 17.80 |
| | Std | 11.28 | 4.60 | 38.38 | 0.95 | 7.74 | 4.54 | 7.92 | 1.12 | 7.42 | 0.25 | 12.07 | 4.60 | 24.96 | 0.87 | 7.64 | 1.21 |
| | N | 2,772 | 2,772 | 2,770 | 2,771 | 2,771 | 2,752 | 2,753 | 2,736 | 2,772 | 793 | 784 | 794 | 794 | 794 | 794 | 793 |
| G2Training | Mean | 82.39 | 57.48 | 1388 | 7.50 | 71.37 | 46.72 | 74.41 | 15.12 | 156.34 | 1.97 | 80.55 | 62.22 | 1,400 | 7.21 | 66.87 | 17.78 |
| | Std | 11.30 | 4.76 | 39.88 | 1.02 | 8.19 | 5.13 | 7.68 | 1.13 | 9.89 | 0.23 | 12.11 | 4.50 | 40.60 | 0.91 | 9.28 | 1.31 |
| | N | 2,965 | 2,964 | 2,964 | 2,963 | 2,964 | 2,951 | 2,952 | 2,958 | 2,964 | 781 | 778 | 782 | 782 | 782 | 782 | 781 |
| G3Training | Mean | 84.85 | 57.92 | 1495 | 7.41 | 76.11 | 47.33 | 75.43 | 15.31 | 159.81 | 1.95 | 82.36 | 63.52 | 1,509 | 7.19 | 72.89 | 18.14 |
| | Std | 9.77 | 4.85 | 42.52 | 1.03 | 7.52 | 4.64 | 7.85 | 1.15 | 6.21 | 0.25 | 11.00 | 4.66 | 36.38 | 0.90 | 7.90 | 1.35 |
| | N | 2,117 | 2,117 | 2,115 | 2,116 | 2,117 | 2,103 | 2,103 | 2,115 | 2,117 | 759 | 755 | 768 | 769 | 769 | 769 | 768 |
| G4Training | Mean | 83.32 | 57.20 | 1460 | 7.37 | 77.15 | 45.22 | 78.10 | 15.10 | 147.57 | 1.77 | 80.02 | 62.65 | 1,496 | 6.87 | 70.93 | 18.09 |
| | Std | 10.28 | 4.92 | 42.79 | 0.98 | 7.72 | 4.74 | 7.86 | 1.23 | 7.82 | 0.27 | 11.02 | 4.77 | 36.61 | 0.94 | 8.59 | 1.38 |
| | N | 2,167 | 2,167 | 2,164 | 2,167 | 2,167 | 2,157 | 2,158 | 2,164 | 2,167 | 768 | 769 | 772 | 772 | 771 | 772 | 769 |
| G5Training | Mean | 85.99 | 58.59 | 1486 | 8.06 | 78.70 | 47.38 | 79.38 | 15.20 | 155.33 | 1.81 | 82.90 | 62.66 | 1,477 | 7.65 | 72.71 | 17.88 |
| | Std | 9.55 | 4.93 | 46.84 | 1.01 | 8.16 | 4.96 | 7.59 | 1.20 | 8.80 | 0.25 | 10.01 | 4.67 | 36.53 | 0.89 | 9.08 | 1.41 |
| | N | 290 | 290 | 289 | 290 | 290 | 278 | 278 | 290 | 290 | 277 | 274 | 280 | 280 | 280 | 280 | 275 |
| G5Validation | Mean | 83.09 | 59.17 | 1,493 | 7.70 | 78.06 | 45.02 | 80.19 | 15.38 | 148.89 | 1.80 | 77.38 | 63.31 | 1,488 | 7.47 | 71.55 | 17.92 |
| | Std | 9.20 | 4.78 | 41.74 | 1.09 | 7.29 | 4.53 | 7.56 | 1.10 | 7.84 | 0.27 | 11.70 | 4.93 | 35.01 | 0.93 | 8.58 | 1.38 |

Early (e) traits recorded at 26-28 weeks of life: egg production (ePD); age at sexual maturity (eSM); shell quality (ePS); weight of first 3 eggs (eE3); color of first 3 eggs (eC3); egg weight (eEW); albumen height (eAH); egg color (eCO); and yolk weight (eYW); late (l) traits recorded at 42-46 weeks: body weight (lBW); egg production (lPD); egg weight (lEW); albumen height (lAH); egg color (lCO); and yolk weight (lYW).

**a** is the ($p$x1) vector of breeding values of genotyped individuals, with variance-covariance matrix $G\ \sigma_a^2$,

**P** is the ($N \times p$) matrix with element $ij = 1$ if the $i$th observation is on genotyped individual $j$, zero otherwise,

**Q** is an ($N \times N$) diagonal matrix with element $ii = 1$ if observation $i$ is on a non-genotyped individual, zero otherwise,

**S** and **D** are ($N \times p$) incidence matrices with elements in rows for non-genotyped individuals that correspond to the columns identifying sires and dams set to 1, and zero's elsewhere.

**e** is the (Nx1) vector of random errors which has variance $\sigma_e^2$ for observations on genotyped individuals and $\sigma_e^2 + \frac{1}{2}\sigma_a^2$ for observations on non-genotyped individuals, ignoring the effect of parental inbreeding on Mendelian sampling variance in progeny.

Population size and avoiding the mating of close relatives insured low inbreeding in this population. Furthermore, variance component estimates from a full animal model and the reduced animal model described above, using pedigree relationships, were very close. Thus, ignoring the effect of parental inbreeding on Mendelian sampling variance in progeny is expected to have a negligible impact on results.

Three models were used to predict breeding values of individuals in generation 5:

1) PBLUP - Reduced animal model using pedigree relationships.

2) GBLUP - Reduced animal model using marker-based relationships for genotyped birds, with covariance matrix derived by the method of VanRaden [11], using allele frequencies based on all genotyped animals.

3) Bayes-C-π - A genomic prediction method similar to Bayes-B of Meuwissen et al. [1], except for the estimation of the proportion of SNP with zero effects (π) and assuming a common variance for all fitted SNP, with a scaled inverse chi-square prior with $v_a$ degrees of freedom and scale parameter S2a, as described by Habier et al. [12]. The prior for π was uniform (0,1). The chain length was 160,000 iterations, with the first 50,000 excluded as the burn in period. In this analysis, the average

genotype (number of 'B' vs. 'A' alleles) of the genotyped parents was used to fit SNP genotype effects to the pre-adjusted mean performance of their non-genotyped progeny. To account for different residual variances for progeny means, residual variances were scaled using weights derived from $w_p = \frac{1-h^2}{(1-0.5h^2)/p}$, where $p$ is the number of phenotypes included in the mean [5].

All models included the fixed effect of hatch within generation, either fitting it in the model (for PBLUP and GBLUP) or pre-adjusting the data by subtracting solutions from a single trait animal model that included all observations and pedigree relationships (for Bayes-C-π). The PBLUP and GBLUP analyses were performed using ASREML [13] and Bayes-C-π using GenSel [12]. The correlation between EBV with hatch-corrected phenotype (as described above) in the validation data sets divided by square root of heritability and regression of hatch-corrected phenotype on EBV were used as measures of accuracy and bias of EBV, respectively. Another comparison of methods was based on selecting the top 30 individuals from the 290 available for validation based on EBV for each trait and comparing the average hatch-corrected phenotype of the selected individuals. Marker based parental average (PA) EBV were also calculated for animals in the validation sets to evaluate the extent to which improvements in accuracy with use of markers resulted from more accurate estimates of Mendelian sampling terms versus more accurate EBV of the parents. This was possible in this population because parents of both sexes were genotyped. To check if combining marker-based estimates with PA increases accuracies of estimates, as suggested by VanRaden et al. [6] for dairy cattle, linear regression of pre-adjusted phenotypes on PA and genomic EBV was performed; if GEBV capture all pedigree information, then adding PA to the regression model is not expected to increase the ability to predict phenotype in validation animals.

## 3. RESULTS AND DISCUSSION

Estimates of heritability from single-trait pedigree-based animal models fitted to the whole data set are shown in Table 2. Estimates were low to moderate for production and shell quality and moderate to high for all other egg quality traits, as expected. Estimates of heritability for early traits were higher than for the corresponding late traits. Variance components for the late traits may be biased because only selected birds had the opportunity to obtain phenotypes for these traits.

**Table 2** Estimates of heritability from single-trait pedigree-based animal model analyses for early (e) traits recorded at 26-28 weeks of life and for late (l) traits recorded at 42-46 weeks

| | | | | | | | | |
|---|---|---|---|---|---|---|---|---|
| | | | | **Early traits** | | | | |
| Trait | ePD | eEW | ePS | eAH | eCO | eYW | eE3 | eC3 |
| $h^2$ | 0.39 | 0.74 | 0.29 | 0.55 | 0.72 | 0.47 | 0.64 | 0.66 |
| | | | | **Late traits** | | | | |
| Trait | lPD | lEW | lPS | lAH | lCO | lYW | lBW | eSM |
| $h^2$ | 0.26 | 0.67 | 0.25 | 0.52 | 0.67 | 0.50 | 0.48 | 0.55 |

[1] Standard errors of heritability were between 0.02 and 0.03; early (e) traits recorded at 26-28 weeks of life: egg production (ePD); age at sexual maturity (eSM); shell quality (ePS); weight of first 3 eggs (eE3); color of first 3 eggs (eC3); egg weight (eEW); albumen height (eAH); egg color (eCO); and yolk weight (eYW); late (l) traits recorded at 42-46 weeks: body weight (lBW); egg production (lPD); egg weight (lEW); albumen height (lAH); egg color (lCO); and yolk weight (lYW).

## 3.1 Accuracy of marker-based EBV

Marker-based EBV had, in general, a higher predictive ability than estimates using pedigree relationships (Figures 1 and 2) for all traits and for early and late selection scenarios. The advantage of GBLUP over PBLUP is due to the fact that realized marker-based genetic similarity between animals deviate from pedigree-based relationship coefficients. In addition, marker-based EBV are not affected by pedigree errors, although they are affected by genotyping errors and errors in DNA sample identification. As shown in Figure 3, marker-based relationships varied substantially around pedigree relationships. The regression of marker-based on pedigree-based relationships was 0.88 for all individuals and 0.97 for validation individuals, demonstrating on average good agreement between both types of relationships. The correlation between the two relationship measures was 0.68 and 0.72 for all and validation individuals, respectively.

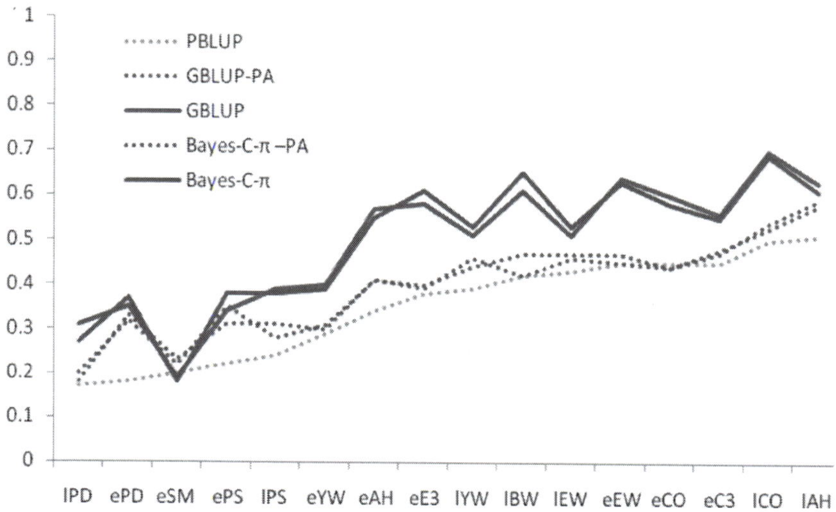


**Figure 1** Accuracy of predicted breeding values and parental average (PA) breeding values from three methods: pedigree-based BLUP (PBLUP), marker-based BLUP (GBLUP), and Bayesian variable selection prediction (Bayes-C-π) in the early selection scenario Accuracy is the correlation between predicted breeding values and hatch-corrected phenotype in the validation set divided by square root of heritability from Table 2.

The difference in accuracy between GBLUP and PBLUP was smaller for selection at a later age than at an early age, when data on sibs of selection candidates were available (Figures 1 and 2). This extra information increased the accuracy of all methods and particularly of PBLUP. Using marker-based relationships increased accuracies up to over two-fold for early selection and by up to 88% for late selection. Proportionally, the highest gain in accuracy was achieved for traits with the lowest heritability. Accuracies obtained with GBLUP were on average slightly larger than those with Bayes-C-π. Several simulation studies have shown that the accuracy of Bayesian methods is higher than that of GBLUP [1,14,15] but a simulation study reported by Daetwyler et al. [16] has shown that the relative performance of GBLUP depends to a large extent on the genetic architecture of the trait. Also, studies on real data in dairy cattle have shown that GBLUP can be equally accurate or even superior in prediction for traits for which no individual QTL explains a large proportion of the variation [17,18].

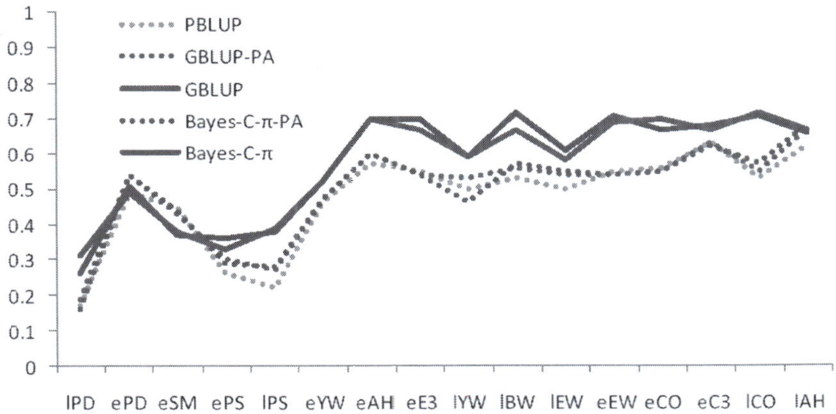


**Figure 2** Accuracy of predicted breeding values and parental average (PA) breeding values from three methods: pedigree-based BLUP (PBLUP), marker-based BLUP (GBLUP), and Bayesian variable selection prediction (Bayes-C-π) in the late selection scenario.. Accuracy is the correlation between predicted breeding values and hatch-corrected phenotype in the validation set divided by square root of heritability from Table 2.

Correlations for selection at an early age between EBV obtained by PBLUP and GBLUP ranged from 0.48 to 0.70 across the traits; from 0.46 to 0.71 between EBV from PBLUP and Bayes-C-π; and from 0.79 to 0.97 between EBV from GBLUP and Bayes-C-π. This indicates that reranking of top individuals is very likely between pedigree- and marker-based methods but limited between GBLUP and Bayes-C-π. This was confirmed by the average performance of the top 30 individuals selected with different methods (Table 3), which was similar for marker-based methods but somewhat different for the group selected based on pedigree EBV. A similar tendency was observed for ranking at late selection but correlations between EBV from different methods were higher for this scenario.

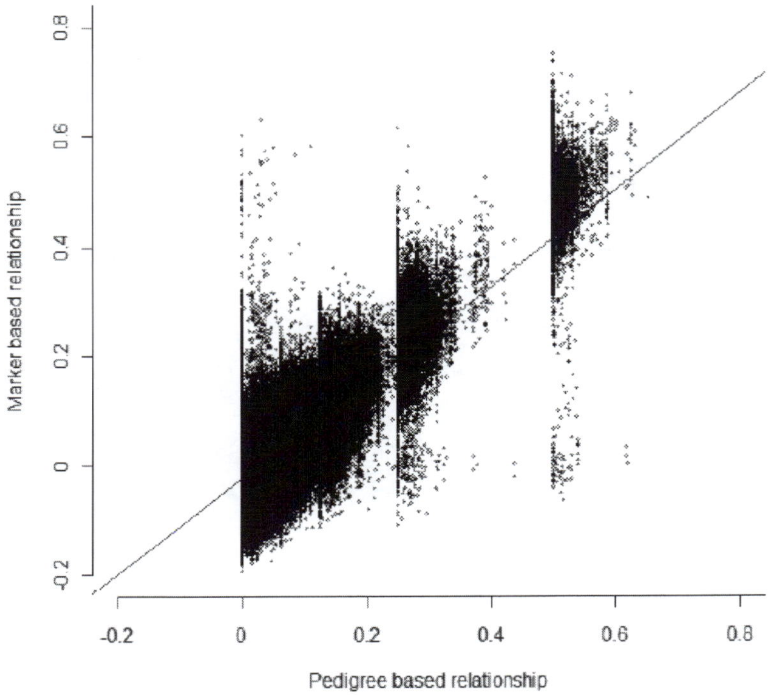


**Figure 3** Pedigree and marker based relationships in the studied population.

The presence of bias in EBV was evaluated by regressing phenotypes of validation individuals on their EBV. On average, these regression coefficients tended to be lower than the expected value of 1: 0.9 for PBLUP, 0.8 for GBLUP and 0.86 for Bayes-C-π, which suggests that EBV overestimated differences in phenotypes of progeny. This bias may be due to selection not being properly accounted for by the single-trait analyses or due to the assumption of normality for genotypic values not being valid. However, the mean squared deviation of the regression coefficients from 1 was lowest for Bayes-C-π, 0.05, compared to 0.06 for PBLUP and 0.07 for GBLUP, suggesting that estimates from Bayes-C- π were least biased. Sib information tended to improve the performance of all methods in this regard for most traits.

**Table 3** Validation of predicted breeding values and parental average (PA) breeding values from three methods: pedigree-based BLUP (PBLUP), marker-based BLUP (GBLUP), and Bayesian variable selection prediction (Bayes-C-π), for early and late selection

| Method | ePD | eEW | ePS | eAH | eCO | eE3 | eC3 | eYW | eSM[1] | IBW[1] | IPD | IEW | IPS | IAH | ICO | IYW |
|---|---|---|---|---|---|---|---|---|---|---|---|---|---|---|---|---|
| EARLY SELECTION | | | | | | | | | | | | | | | | |
| Slope from regression of phenotype on EBV | | | | | | | | | | | | | | | | |
| PBLUP | 0.63 | 1.12 | 0.71 | 0.87 | 0.93 | 0.88 | 0.85 | 0.70 | 0.56 | 1.06 | 0.52 | 1.05 | 0.56 | 1.16 | 1.03 | 0.88 |
| GBLUP | 0.53 | 0.87 | 0.58 | 0.93 | 0.70 | 0.81 | 0.67 | 0.58 | 0.34 | 1.07 | 0.54 | 0.73 | 0.61 | 1.05 | 0.93 | 0.82 |
| Bayes-C-π | 0.65 | 0.93 | 0.68 | 0.94 | 0.69 | 0.86 | 0.73 | 0.59 | 0.34 | 1.01 | 0.56 | 0.91 | 0.72 | 1.13 | 0.98 | 0.91 |
| Average performance of top 30 individuals | | | | | | | | | | | | | | | | |
| PBLUP | 89.9 | 61.0 | 1459.3 | 7.78 | 84.2 | 46.0 | 82.2 | 15.4 | 148.0 | 1.78 | 80.8 | 65.1 | 1453.3 | 7.29 | 76.5 | 18.1 |
| GBLUP | 90.3 | 63.5 | 1452.4 | 8.38 | 86.8 | 47.4 | 83.3 | 15.5 | 147.3 | 1.73 | 79.9 | 65.1 | 1440.9 | 7.22 | 80.0 | 18.3 |
| Bayes-C-π | 91.2 | 62.0 | 1453.7 | 8.41 | 85.9 | 48.3 | 83.4 | 15.4 | 147.2 | 1.70 | 78.1 | 64.7 | 1449.9 | 7.12 | 80.3 | 18.3 |
| LATE SELECTION | | | | | | | | | | | | | | | | |
| Slope from regression of phenotype on EBV | | | | | | | | | | | | | | | | |
| PBLUP | 1.08 | 0.90 | 0.51 | 1.13 | 0.93 | 0.85 | 1.07 | 0.80 | 0.90 | 1.12 | 0.47 | 1.04 | 0.46 | 1.29 | 0.95 | 0.97 |
| GBLUP | 0.72 | 0.85 | 0.50 | 1.06 | 0.82 | 0.83 | 0.81 | 0.72 | 0.60 | 1.10 | 0.54 | 0.83 | 0.60 | 1.06 | 0.91 | 0.89 |
| Bayes-C-π | 0.81 | 0.93 | 0.57 | 1.10 | 0.80 | 0.89 | 0.86 | 0.75 | 0.65 | 1.06 | 0.51 | 1.01 | 0.69 | 1.13 | 0.97 | 0.97 |

[1]low values are desired.

## 3.2 Estimation of π, the proportion of markers with zero effects

The proportion of markers with zero effect (π) is estimated from the data in the Bayes-C-π method. Habier et al. [12] have shown that, if there is enough information in the data, $(1- \hat{\pi})k$ is a good estimate of the number of QTL affecting the trait when k unlinked SNP with normally distributed effects were simulated and genotypes used for training included genotypes at the QTL. In the case of more realistic simulations, where QTL genotypes were not included as markers but the effects were estimated based on k linked markers, the number of markers fitted was higher than the number of true QTL, but the tendency for lower estimates of π for scenarios with more QTL did hold [12].

The posterior means of π (Table 4) suggest that a high proportion of markers should be included in the model to explain a substantial part of the genetic variation for the majority of traits in our data; estimates of π ranged from 0.19 to 0.99, which suggests that between 111 and 19,541 markers explained variation for the analyzed traits (Table 4). The large number of associated markers with relatively small effects explains the good performance of GBLUP, which assumes a polygenic determination of traits. However, GBLUP also performed well for egg weight and body weight, which had very high estimates of π. The results suggest that a limited number of markers explain most of the genetic variation for body size in chickens. This can be due to these markers being linked to or in linkage disequilibrium with QTL and/or due to markers capturing pedigree relationships [19].

The accuracy of estimates of π depends on the information content of the data and on mixing in the Monte Carlo Markov Chain, which can be poor for Bayes-C- π. Two independent chains with a high (0.99) or a low (0.1) starting value for π were used to verify convergence of π. For some traits (eE3, eEW, lEW, eCO, lBW), both chains converged to the same value with a clearly peaked posterior distribution but for other traits 160,000 iterations were not sufficient for the two chains to reach the same posterior means, as the posterior distribution of π was relatively flat. This difference may reflect differences in genetic architecture of the traits. For traits with a high estimate of π (i.e. with few markers associated), convergence was obtained quickly and the standard deviation of the posterior distribution of π was small but for traits for which many markers were fitted in the model, the standard deviation of π was high, which suggests that models with different numbers of markers had similar likelihoods. Nevertheless, lack of convergence in π, i.e. different estimates depending on starting value, had almost no impact on the accuracy of EBV. There was also no substantial difference between early and late selection scenarios with regard to convergence of estimates of π. Only for ePD and lCO did the inclusion of additional information from sibs make the posterior means of π from different chains more similar to each other.

**Table 4** Estimates of the proportion of markers with zero effects (x100 ± SD) from the Bayesian variable selection model with starting values of 0.1 (π = 0.1) or 0.99 (π = 0.99)

| | | | | | | | | |
|---|---|---|---|---|---|---|---|---|
| | | | | **Early selection** | | | | |
| Trait | ePD | eEW | ePS | eAH | eCO | eYW | eE3 | eC3 |
| π = 0.1 ± SD | 34 ± 19 | 98 ± 0 | 33 ± 20 | 42 ± 32 | 90 ± 5 | 60 ± 29 | 98 ± 0 | 48 ± 29 |
| π = 0.99 ± SD | 21 ± 21 | 98 ± 0 | 58 ± 27 | 71 ± 20 | 91 ± 3 | 45 ± 25 | 98 ± 1 | 60 ± 26 |
| Trait | lPD | lEW | lPS | lAH | lCO | lYW | lBW | eSM |
| π = 0.1 ± SD | 19 ± 17 | 99 ± 0 | 42 ± 27 | 37 ± 31 | 38 ± 23 | 36 ± 16 | 99 ± 3 | 33 ± 24 |
| π = 0.99 ± SD | 34 ± 25 | 99 ± 0 | 49 ± 30 | 30 ± 21 | 56 ± 30 | 90 ± 9 | 99 ± 3 | 58 ± 27 |
| | | | | **Late selection** | | | | |
| Trait | ePD | eEW | ePS | eAH | eCO | eYW | eE3 | eC3 |
| π = 0.1 ± SD | 38 ± 28 | 98 ± 0 | 40 ± 22 | 82 ± 9 | 92 ± 4 | 64 ± 24 | 97 ± 1 | 69 ± 16 |
| π = 0.99 ± SD | 36 ± 21 | 98 ± 1 | 35 ± 22 | 51 ± 29 | 92 ± 3 | 39 ± 20 | 97 ± 1 | 52 ± 33 |
| Trait | lPD | lEW | lPS | lAH | lCO | lYW | lBW | eSM |
| π = 0.1 ± SD | 36 ± 17 | 98 ± 0 | 41 ± 24 | 59 ± 26 | 40 ± 22 | 43 ± 32 | 99 ± 2 | 48 ± 23 |
| π = 0.99 ± SD | 49 ± 31 | 98 ± 1 | 24 ± 21 | 41 ± 27 | 45 ± 20 | 57 ± 28 | 99 ± 2 | 32 ± 19 |

## 3.3 Information from parental average EBV

In dairy cattle, genomic predictions are often combined with pedigree information [4] before obtaining final genomic EBV. In our study, lEW was the only trait for which adding pedigree-based information significantly improved predictive ability. The increase in the R-square of the regression equation to predict hatch-corrected phenotypes from generation 5 when adding PBLUP to marker-based EBV was significant ($p < 0.05$) only for lEW for GBLUP and Bayes-C-π, for which the R-square increased from 0.174 to 0.189 and from 0.187 to 0.203, respectively. Increases in R-square were not significant ($p > 0.05$) for all other traits using both methods. This suggests that in this dataset, the markers capture most of the pedigree information, likely because all the parents were genotyped.

For most traits, the predictive ability of the marker-based EBV was not substantially lower for traits measured at a late age (Figures 1 and 2), although late traits were only recorded on selected individuals and estimated heritabilities for late traits were generally lower than for the corresponding traits measured at a younger age. This indicates that having records only on selected parents did not limit the ability to estimate marker effects.

In Figures 1 and 2, the difference between the accuracy of marker- versus pedigree-based parental average EBV (e.g. GBLUP-PA vs. PBLUP) reflects the gain in information from more accurate EBV of parents when using markers, while the difference between the accuracy of marker-based parental average EBV and marker-based individual EBV (e.g. GBLUP-PA vs. GBLUP) arises from markers providing information on Mendelian sampling terms. For ePS and ePD and eSM, the increase in accuracy at an early age could be attributed mostly to better estimates of parental EBV. For all other traits, increases in accuracy were primarily based on markers providing information on Mendelian sampling terms. For EBV for selection at a later age, the improvement originated mostly from Mendelian sampling terms, probably because the pedigree parental average EBV were much more accurate than at the earlier age.

## 3.4 Reduced animal model

The reduced animal model was used to incorporate genomic information into genetic evaluation using GBLUP. It was possible to use this model here because all the parents were genotyped, thus data from non-genotyped individuals could be included without loss of information. If some parents are not genotyped, the 1-step methods that combine pedigree-based and genomic relationships can be used to avoid loss of information [7,8]. An alternative to the 1-step method is the use of deregressed EBV [5,6] but this involves approximations and a potential loss of information.

In fact, the model used here represents a special case of the 1-step method of Legarra et al. [7], where all non-genotyped individuals in the data are non-parent progeny. In this case, the only pedigree-relationships that are used are those between genotyped parents and their non-genotyped progeny. Without inbreeding, the expectation of these relationships is equal to 0.5, both based on pedigree and based on genomic data, because progeny receive half of their alleles from each parent. Thus, in this special case, combining genomic and pedigree relationships does not require the rescaling that is typically required for the 1-step approach [20]. In addition to avoiding the need for rescaling, this special case allows equations for non-parents to be absorbed, as in the reduced animal model, which reduces computational demands, although the main computational task of inverting the dense genomic relationship matrix of genotyped individuals remains. By absorbing non-parents, computing time for the reduced animal model is proportional to $n^3$, where n is the number of genotyped animals, while the number of animals with phenotypes has a negligible impact on computing time. Computing time for Bayes-C-π is proportional to the number of markers and to the number of records. The reduced animal model can also easily be extended to a multi-trait setting, following standard multiple-trait animal model procedures. Finally, applying a reduced animal model makes it possible to use weighted genomic relationship matrices that accommodate differential weights on SNP, depending on their effects, similar to the Bayesian model averaging methods [21]. Use of a weighted genomic relationship matrix in a multi-trait setting, however, requires further work.

## 3.5 Implementation of genomic selection in layer chickens

Increases in accuracy were evaluated when selection is at a very early age, prior to phenotypes being available on selection candidates or their sibs, and at a later age. Late age selection represents a scenario in which genomic information is used to increase accuracy of selection in existing layer breeding programs, particularly in the case of males, which are primarily evaluated based on sib information in current breeding programs. Early age selection represents a scenario in which the benefits of genomic selection are capitalized on by also reducing the generation interval from the traditional one year to half a year, as proposed by Dekkers et al. [22]. Using these results, breeding programs exploiting genomic information can be optimized, including scenarios where only male candidates are genotyped and where population sizes are reduced to capitalize on the effect of GEBV on rates of inbreeding. The use of low-density SNP panels needs to be evaluated [23] to reduce costs of genotyping, but this was beyond the scope of this research. In this study, the size of the training data was limited compared to what is available in dairy cattle and increasing its size is expected to further increase the accuracy of GEBV.

## 4. CONCLUSIONS

Reduced animal model approaches can be used to estimate breeding values from high-density SNP data when all parents have been genotyped. Marker-based methods improve the prediction of future performances compared to the classical pedigree-based approach, with most of the accuracy increase due to improved estimation of Mendelian sampling terms. The advantage of marker-based methods is greater for selection at a young age, before information on sibs of selection candidates is available. The accuracies of methods that assume equal variance for all SNP, such as GBLUP and of those that allow differential weighting and shrinkage of SNP effects are similar.

## COMPETING INTERESTS

The authors declare that they have no competing interests.

## AUTHORS' CONTRIBUTIONS

All authors conceived the study, contributed to methods and to writing the paper and also read and approved the final manuscript. AW undertook the analysis and wrote the first draft. Data were prepared by JA, PS, JF and NPO. JCMD provided overall oversight of the project.

## ACKNOWLEDGEMENTS

This study was supported by Hy-Line Int., the EW group, and Agriculture and Food Research Initiative competitive grants 2009-35205-05100 and 2010-65205-20341 from the USDA National Institute of Food and Agriculture Animal Genome Program. Ian White helped with the REML analysis.

## REFERENCES

1 Meuwissen THE, Hayes BJ, Goddard ME. Prediction of total genetic value using genome-wide dense marker maps. Genetics. 2001; 157:1819–1829.

2 Daetwyler HD, Villanueva B, Bijma P, Woolliams JA. Inbreeding in genome-wide selection. J Anim Breed Genet. 2007;124:369–376.

3 Goddard M. Genomic selection: prediction of accuracy and maximisation of long term response.Genetica. 2009;136:245–257.

4 Hayes BJ, Bowman PJ, Chamberlain AJ, Goddard ME. Genomic selection in dairy cattle: Progress and challenges. J Dairy Sci. 2009;92:433–443.

5 Garrick DJ, Taylor JF, Fernando RL. Deregressing estimated breeding values and weighting information for genomic regression analyses. Genet Sel Evol. 2009;41:55.

6 VanRaden PM, Van Tassell CP, Wiggans GR, Sonstegard TS, Schnabel RD, Taylor JF, Schenkel FS. Reliability of genomic predictions for North American Holstein bulls. J Dairy Sci. 2009;92:16–24.

7 Legarra A, Aguilar I, Misztal I. A relationship matrix including full pedigree and genomic information.J Dairy Sci. 2009;92:4656–4663.

8 Christensen OF, Lund M. Genomic prediction when some animals are not genotyped. Genet Sel Evol.2010;42:2.

9 Quaas RL, Pollak EJ. Mixed model methodology for farm and ranch beef cattle testing programs. J Anim Sci. 1980;51:1277–1287.

10 White IMS, Roehe R, Knap PW, Brotherstone S. Variance components for survival of piglets at farrowing using a reduced animal model. Genet Sel Evol. 2006;38:359–370.

11 VanRaden PM. Efficient methods to compute genomic predictions. J Dairy Sci. 2008;91:4414–4423.

12 Habier D, Fernando RL, Kizilkaya K, Garrick DJ. In: Proceedings of the 9th World Congress on Genetics Applied to Livestock Production. Leipzig, editor. German Society for Animal Science; 2010. Extension of the Bayesian Alphabet for Genomic Selection; p. 468. 1-6 August 2010.

13 Gilmour AR, Gogel BJ, Cullis BR, Thompson R. ASReml User Guide Release 3.0. VSN Int Ltd.2008.

14 Lund MS, Sahana G, de Koning DJ, Su G, Carlborg O. Comparison of analyses of the QTLMAS XII common dataset. I: Genomic selection. BMC Proc. 2009;3(Suppl 1):566.

15 Bastiaansen JWM, Bink MCA, Coster A, Maliepaard C, Calus MPL. Comparison of analyses of the QTLMAS XIII common dataset. I: genomic selection. BMC Proc. 2010;4(Suppl 1):S1.

16 Daetwyler HD, Pong-Wong R, Villanueva B, Woolliams JA. The impact of genetic architecture on genome-wide evaluation methods. Genetics. 2010;185(3):1021–31.

17 Luan T, Woolliams JA, Lien S, Kent M, Svendsen M, Meuwissen THE. The accuracy of genomic selection in Norwegian Red Cattle assessed by cross-validation. Genetics. 2009;183:1119–1126.

18 Verbyla KL, Hayes BJ, Bowman PJ, Goddard ME. Accuracy of genomic selection using stochastic search variable selection in Australian Holstein Friesian dairy cattle. Genet Res. 2009;91:307–311.

19 Habier D, Fernando RL, Dekkers JCM. The impact of genetic relationship information on genome-assisted breeding values. Genetics. 2007;177:2389–2397.

20 Aguilar I, Misztal I, Johnson D, Legarra A, Tsuruta S, Lawlor T. A unified approach to utilize phenotypic, full pedigree, and genomic information for genetic evaluation of Holstein final score. J Dairy Sci. 2010;93:743–752.

21 VanRaden PM. Genomic measures of relationship and inbreeding. Interbull Bull. 2007;37:33–36.

22 Dekkers JCM, Zhao HH, Habier D, Fernando RL. Opportunities for genomic selection with redesign of breeding programs. J Anim Sci. 2009;87(Suppl E):275.

23 Habier D, Fernando RL, Dekkers JCM. Genomic selection using low-density marker panels. Genetics.2009;182:343–353.

# CHAPTER 9

# Association between UCP Gene Polymorphisms and Growth, Breeding Value of Growth and Reproductive Traits in Mazandaran Indigenous Chicken

**Seyed Mohammad Moazeni[1], Mohammadreza Mohammadabadi[2], Mostafa Sadeghi[3], Hosein Moradi Shahrbabak[3], Ali Esmailizadeh Koshkoieh[2], Farhad Bordbar[1]**

[1]MSc Student of Department of Animal Science, Shahid Bahonar University of Kerman, Kerman, Iran

[2]Associate Professor of Department of Animal Science, Shahid Bahonar University of Kerman, Kerman, Iran

[3]Assistant Professor of Department of Animal Science, Tehran University, Tehran, Iran

## ABSTRACT

The objective of this study was to estimate the allele and genotype frequencies of the UCP/Hha1 polymorphisms and to determine associations between these polymorphisms and growth traits, breeding value of growth and reproductive traits in Mazandaran Indigenous Chicken. To estimate breeding value of the reproductive traits, and phenotypic information of 18 successive generations of selection in breeding station of Mazandaran Indigenous Chicken (north of Iran), 190 birds were analyzed using a univariate animal model in ASREML procedure.. Genotyping of UCP gene polymorphism for all subjects was performed by the PCR?RFLP method. The evaluation of the association between this SNP with reproductive traits suggests a positive effect of TC genotype with age at first egg (ASM) compared with CC genotype. In addition, TC genotype was significantly associated with breeding value of age at first egg compared with CC genotype ($P < 0.05$). In conclusion, our results suggest that the TC genotype of the UCP gene is associated with age at sexual maturity (ASM) and breeding value of age at sexual maturity, and UCP polymorphisms may be used as DNA markers for selection in the breeding process of Mazandaran Indigenous Chicken.

## KEYWORDS

Reproductive Traits, Polymorphism, Chicken, UCP, PCR-RFLP

## 1. INTRODUCTION

Low energy expenditure, in addition to increased energy intake, has been a major cause of future weight gain, and variations in energy expenditure may be one of the underlying sources of variation in body weight [1] . Various determinants, including body composition, hormonal levels, activity of the sympathetic nervous system, and genetics are responsible for differences in metabolic rate among individuals [2] . A mitochondrial protein called uncoupling protein plays an important role in generating heat and burning calories by creating a pathway that allows dissipation of the proton electrochemical gradient across the inner mitochondrial membrane in brown adipose tissue, without coupling to any other energy-consuming process. This pathway has been implicated in the regulation of body temperature, body composition, and glucose metabolism [3] . The uncoupling protein (UCP) gene is one of the many genes that maybe related with growth traits, breeding value of growth and reproductive traits. UCP3 is the member of uncoupling proteins (UCPs) family of mitochondrial transporters, all of which are known to uncouple oxidative phosphorylation via proton leakage from the inner mitochondrial membrane. Thermoregulation and energy metabolism are potentially regulated by these mitochondrial proteins [4] . The uncoupling protein gene (UCP) is a candidate gene for growth traits, breeding value of growth and reproductive traits because it may contribute to the proton leak that occurs in mitochondria and partially uncouples the oxidative phosphorylation, thus releasing chemical energy as heat [5] . Uncoupling protein (UCP) gene family involved in energy metabolism of body, and had significant effects on energy balance related traits, such as body weight resting metabolic rate, and efficiency of food conversion etc. [6] . Liu et al. [7] found UCP-3 gene polymorphisms may contribute to body mass index (BMI) in the Caucasian population. Rudofsky et al. [8] found that human UCP-3 gene involved in energy balance and was associated with obesity and diabetes. UCP2 gene is expressed extensively in various tissues, like white adipose tissue, pancreas, skeletal muscle, and liver, in human [9] . Both environmental and genetic factors may have contribution toward susceptibility of growth traits, breeding value of growth and reproductive traits. Genetic diversity in indigenous breeds is a major concern considering the necessity of preserving what may be a precious and irreplaceable richness, regarding new productive demands. Conservation should be based on a deep knowledge of the genetic resources of the specific breed. Therefore, it is important to try to characterize genetically indigenous breeds [10]. A species without enough genetic diversity is thought to be unable to cope with changing environments or evolving competitors and parasites and also the ability of a population to respond adaptively to environmental changes depends on the level of genetic variability or diversity it contains. Therefore, studies of population genetic diversity and the structure of population within and between species may not only illustrate the evolutionary process and mechanism but also provide information useful for biological conservation of the Bovidae family, sheep, goats, birds and so on [11] . Molecular markers are increasingly used for the study of genetic diversity of populations in recent years [12] . Therefore, in the present study, we aimed to investigate association of polymorphism of UCP2 gene with growth traits, breeding value of growth and reproductive traits in subjects from Iran.

# 2. MATERIAL AND METHODS

## 2.1. Experimental Population and Sampling

Breeding station of Mazandaran Indigenous Chicken is located at 28 km far from Sari, the provincial capital of Mazandaran state, located in the north of Iran . In 1986 about 5000 cocks and hens were purchased from rural regions across the Mazandaran province and kept in a quarantine farm for a year. From those, about 2500 birds of two sexes were kept to produce hatching eggs and chicks produced from these eggs were transferred to the station in 1988. Since then birds have been individually tagged and trap nest has been used for pedigree recording. Parents of each generation (about 100 cocks and 800 hens) are selected among 6000 pedigreed and performance recorded birds produced each generation. In August 2009, a total of 205 blood samples from Mazandaran Indigenous Chicken, including 10 males and 195 females, were collected. Individuals were reared in Native chicken breeding station of Mazandaran and they belonged to generation 17 of the breeding station pedigreed animals. Individuals of this generation were developed by crossing 80 sires and 751 dams from generation16. Approximate 1 ml blood per chick from plumage veins was collected and kept in a tube containing anticoagulant EDTA (Ethylenediaminetetraacetic acid). All samples were delivered back to the laboratory in an ice box. The genomic DNA was extracted from white blood cells using standard salting out procedure described by Abadi et al. [13] . The DNA samples were dissolved in TE buffer which was made from 10 mMTris-Cl (pH 7.5) and 1 mM EDTA (pH 8.0) and were stored at 20°C for use.

## 2.2. Primer Synthesis and PCR-RFLP Reactions

The primers were designed on the basis of DNA sequence of the UCP (Accession: AF433170) using the oligonucleotide design tool Primer 5.0 software (F: 5'-TACCCCAAGCATGCAGAACTCA-3' and R: 5'-

GGAACCGCACCTTGACCAC-3'). Both PCR reactions were performed in a 20 µl mixture containing 10 pmol primers, 200 lMdNTP (deoxyribonucleotide triphosphate), 2 µl 10 × reaction buffer which contained 1.5 mM $MgCl_2$, 1 unit of Taq-DNA polymerase (Promega, Madison, WI), and 50 ng genomic DNA as template. PCR method was used to optimize the reaction accuracy: 94°C for 5 min, 35 cycles of 94°C for 30 s, annealing at 55°C for 60 s, 72°C for 60 s, and a final extension at 72°C for 7 min. PCR products were electrophoretically separated on 2% agarose gel (5 V/cm), stained with ethidium bromide. PCR products were digested by 10 units of Hha1 restriction enzymes (Fermentase, Lithuania ), 6 ml of PCR product, 1.4 ml of Tango buffer and 2 ml nuclease-free water. The final volume of 10 ml was incubated in 37°C for 12h. The fragments were separated a 3.5% agarose gel stained with ethidium bromide.

## 2.3. Traits Measured

Whole information data file (18 successive generations) consisted of three fixed effects (generation, sex and hatch) and 11 recorded traits including body weight at hatch (BW1), body weight at age of 8 (BW8), 12 (BW12) weeks, body weight at sex maturation (WSM), age at first egg (ASM), egg number (EN), first egg weight EW1), average egg weight at age of 28 (EW28), 30 (EW30), 32 (EW32) weeks and average egg weight for the first 12 weeks of production (EW12). BW1, BW8 and BW12 have been measured in both male and female chicken. Also, three combined traits consisting of average of EW28, EW30 and EW32 (AV), intensity of egg production (EINT = (egg number/days recording) × 100) and egg mass (EM = EN × EW12) were calculated to use in the analyses (Table 1). During 18 generations, genetic evaluation of the birds for body weight at 8 weeks, age of the hens at first egg, average egg weight and total number of eggs laid during first 12 weeks after flocks maturity (when 5% of the flock are in egg production) have been performed. Economic indexes are calculated for these traits and birds of two sexes are selected based on their aggregate genotypes for these traits. The goals of the breeding station on the one hand are to increase body weight, egg weight and egg number and on the other hand, to decrease age at first egg.

## 2.4. Statistical Analyses

Pedigree and data file were prepared using Visual FoxPro 9.0 software, the relational data base management system. SAS9.1 package was used to carry out descriptive statistics and fitting model. Fixed effects and their interactions were considered in animal model that provided having significant effect. Genetic analyses were performed using ASREML software [14] . Breeding values of growth and egg production traits was estimated using the BLUP based on an animal model with a relationship matrix. The used models in matrix notation were as follows:

$$Y = Xb + Za + e$$

where, Y is the vector of observations; b the vector of fixed effects of generation, sex and hatch; a the vector of random direct genetic effects; e the vector of random residual effects; X and Z are incidence matrices relating the observations to the respective fixed and direct genetic effects. Calculation of gene frequency was based on direct gene count method by:

$$f(A) = (2nAA + nAa)/2n \text{ or } f(G) = (2nGG + nGg)/2n$$

and standard error of frequency was calculated as $(p(1 - p)/2n)^{1/2}$, where n is the sample size, p is the frequency of A or G allele. Marker-trait association analyses were conducted using following model in GLM procedure of SAS9.1 software. The significant differences of least squares means were tested with Tukey-Kramer's multiple range tests, and a P-value of ≤0.05 was considered statistically significant.

$$f(A) = (2nAA + nAa)/2n \text{ or } f(G) = (2nGG + nGg)/2n$$

$Y_{ijk}$ is the estimated breeding values of traits, μ the population mean, $M_i$ the fix effect of genotypes, $e_{ijk}$ the residual random error. There was no significant interaction between the genes additive effects.

# 3. RESULTS

## 3.1. Analysis of the Polymorphisms in the UCP Gene

Five variations in the UCP gene are shown in Table 1 with data obtained from 190 chickens. A single nonsynonymous SNP was a C > T substitution at nt position 1270 in exon-3 replacing Ala > Val. A PCR-RFLP test using HhaI restriction enzyme was designed to screen the UCP Ala118Val polymorphism (Table 2). Figure 1 shows the electrophoretic gel pattern of the HhaI PCR-RFLP test where lanes 6 and 2 are Ala/Val heterozygotes with or without the 19-bp deletion, respectively. An additional restriction cut site was created within the PCR- amplified region by t > C silent substitution for cysteine at nt position 1316 in exon-3 that produced the second band from the top as seen in Figure 1 for lane 5. The digested allele as allele ‘‘T’’, and the undigested allele was identified as allele ‘‘C’’. Three genotypes TC (Ala/Val), TT (Ala/Ala) and CC (Val/Val), were detected after PCR-RFLP analysis of the UCP gene. The allele frequencies were 0.21 for T and 0.79 for C, and the genotypic frequencies were 0.18, 0.12, and 0.70 for TC, TT and CC, respectively.

## 3.2. Relationship between Polymorphisms and Growth, Breeding Value of Growth and Reproductive Traits

The association between the AF433170: c. C1270T polymorphisms in the UCP gene and growth, breeding value of growth and egg production traits are shown in Table 3 and Table 4, respectively. UCP gene, the Missense mutations were significantly linked to ASM ($P < 0.05$) and also highly significantly linked to breeding value of ASM ($P < 0.05$). In this population, ASM was higher in the heterozygote (UCP Ala/Val) compared with the homozygote (UCP Val/Val).

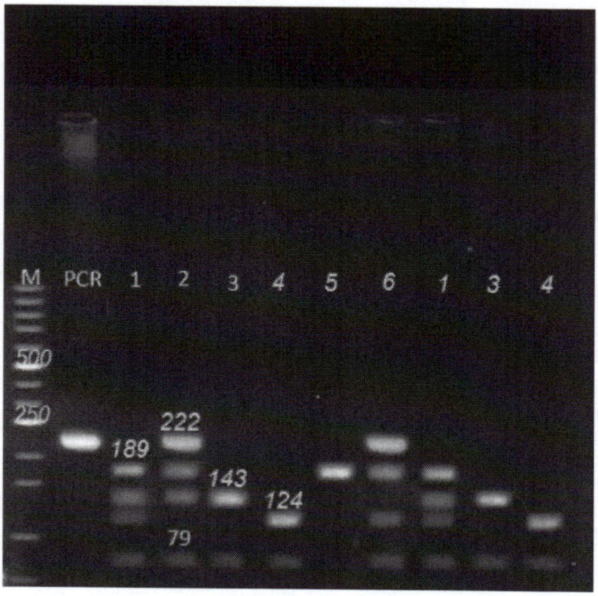


**Figure 1**. The electrophoretic gel patterns of the HhaI PCR- RFLP. M is Ladder, PCR is Fragment of PCR product, lanes 2 and 6 are heterozygote and other lanes are homozygote.

**Table 1**. Statistical description of data set for growth and egg production traits.

| *Traits* | *No. of animal* | *Mean* | *Coefficient of variation* |
|---|---|---|---|
| BW1 (gr) | 35,287 | 35.53 | 8.15 |
| BW8 (gr) | 43,067 | 563.7 | 17.09 |
| BW12 (gr) | 38,297 | 953.9 | 14.49 |
| WSM (gr) | 31,147 | 1694 | 11.90 |
| ASM (day) | 31,349 | 165.5 | 9.23 |
| EN (number) | 31,349 | 36.66 | 39.78 |
| EW1 (gr) | 27,294 | 41.21 | 15.74 |
| EW28 (gr) | 17,225 | 46.91 | 8.48 |
| EW30 (gr) | 19,031 | 48.12 | 8.50 |
| EW32 (gr) | 18,955 | 49.22 | 8.29 |
| EW12 (gr) | 18,847 | 46.62 | 9.28 |
| AV (gr) | 28,725 | 46.84 | 13.05 |
| EM (gr) | 28,725 | 1768 | 39.89 |
| EINT (%) | 31,349 | 57.07 | 33.29 |

BW1, BW8, BW12 = Body weight at birth, 8, 12 weeks of age, WSM = Body weight at sexual maturity, ASM = Age at first egg, EN = Egg number, EW1 = Weight of first egg, EW28, EW30 and EW32 = Average egg weight at 28, 30 and 32 weeks of age respectively, EW12 = Average egg weight for first 12 weeks of production, AV = Average for EW28, 30 and 32, EM = Egg mass (=EN × EW12), EINT = Egg production intensity (= Egg number/days recording) × 100).

**Table 2**. Summary of variations in UCP gene.

| Location | Position | Nucleotide changes | Amino acid changes | Subtype |
|---|---|---|---|---|
| Exon 3 | 1270 | C>T | Ala118Val | Missense |
| | 1316 | T>C | Cys133Cys | Silent |

# 4. DISCUSSION

The study of candidate genes is one of the primary methods to find specific genes related to economic traits in farm animals. UCP involved in regulation of energy metabolism has important physiological effects on economical important traits of farm animals. Gene polymorphisms can be used for improvement of the production traits by genetic selection, if the allelic association with the traits is determined. The chicken UCP gene was first discovered by Raimbault et al. [15] . Its function is still not clear, though there has been some research about it since its discovery. The nucleic acid sequence of avian UCP is highly homologous to both mammalian UCP2 and UCP3 [16] . The current hypothesis on the function of these two UCPs involves regulation of fatty acid oxidation, especially during metabolic stress [17] , possibly by transporting fatty acid ions out of the mitochondria [18] . Two significant associations ($P < 0.05$) were observed with growth, breeding value of growth and reproductive traits in this study (Table 3 and Table 4). Sherman et al. [19] observed significant association between a SNP (A/G) in bovine UCP-3 gene intron-3 and average daily gain and partial efficiency of growth in 9 Continental British hybrid beef steers. Han [20] reported the existence of three SNPs (G/A, C/T and G/T) in exon 3 and four genotypes in Chinese Qinchuan cattle. Association studies showed that the genotype AA had higher slaughter weight, carcass weight, carcass length, eye muscle area, water holding capacity, and marbling than other genotypes; the genotype AB had higher back fat than other genotypes. Li [21] found that the allele A of UCP-3/BglI polymorphism had higher frequencies than allele B in Nanyang, Luxi, and Yanbian cattle. Association analysis showed the superiority of the AA genotype over the AB genotype for the content of b-globin in Luxi cattle. Fasting, which is associated with a major reduction in total body energy expenditure, is associated with an increase in the expression of UCP2 and UCP3 [22] . However, it was also reported that UCP2 and UCP3 genes were unlikely to have a substantial effect on variation in obesity phenotypes, in particular US Caucasian human population [23] . In animal and human studies, UCP2 mRNA levels have been correlated with plasma-free fatty acid concentrations [24] . In adipose tissue of obesity-resistant mice, UCP2 expression is increased 2-fold when compared with obesity-prone animals [25] . Independent studies have identified quantitative trait loci for obesity or body fat in mice that coincide with the genomic location of the UCP2 gene [26] [27] . The SNP in the exon included synonymous and nonsynonymous polymorphisms. At any given position in a DNA sequence, a nucleotide can be substituted by any of the 4 nucleotide bases and may result in biallelic SNP. This occurs due to the low substitution rate of single nucleotides, estimated to be between $1 \times 10^{-9}$ and $5 \times 10^{-9}$ per nucleotide per year at neutral positions in mammals [28] . Based on these numbers, the probability of 2 independent base changes occurring at a single position is very low [28]. A large variation in UCP gene sequences in the studied population was found to reside in the first half of the gene. In the present study, a Missense polymorphism identified in the UCP gene at Ala118Val can be predicted to occur in the transmembrane region of the second domain close to the mitochondrial inner membrane space. The various biological functions of UCP are not known in detail, and based on the localization and nature of the codon UCP118 variant, it is difficult to predict if the encoded protein change may cause defects in the functionality of the UCP. Structural-functional studies are thus needed to elaborate on the effect of the UCP Ala118Val variant on UCP's protein structure as well as function. In a highly selected commercial population, changes in the allele and genotypic frequencies between generations may result from the net reproductive success of the individuals selected as breeding stock with the different genotypes and the selection norms adopted by the breeder. Variable associations of the identified polymorphisms may be a result of the differences in the population characteristics, sex, or both, indicating that the selection criteria may influence the production trait associations. This should be taken into consideration while selecting for the desired reproductive traits. Additional studies are required to expand the genetic, physiological, behavioral aspects involved in feed intake, digestion, and metabolism. The genomic diversity also has important implications in the evolutionary dynamics of species. Investigations of polymorphisms are useful for better understanding of the gene function, and those associated with commercially significant growth, breeding value of growth and reproductive traits have a potential for usage as molecular markers for selection programs. The identified polymorphisms and their associations with the traits of economic importance in present study provide greater insight into the role of gene involved in energy balance in poultry (UCP) and point toward the potential application of the findings for the enhancement of production traits by marker assisted selection.

**Table 3**. Association of the UCP genotypes at the growth and egg production traits (Mean ± S.E.).

| Traits | Genotype TT | Genotype TC | Genotype CC |
|---|---|---|---|
| WSM | 1795.28 ± 35.80 | 1842.07 ± 57.15 | 1721.10 ± 20.80 |
| ASM | $185.19 \pm 3.45^{ab}$ | $189.29 \pm 4.91^{a}$ | $176.08 \pm 1.70^{b}$ |
| EN | 39.97 ± 1.36 | 42.58 ± 2.17 | 40.00 ± 0.79 |
| EW28 | 48.33 ± 0.84 | 48.77 ± 1.46 | 47.86 ± 0.45 |
| EW30 | 49.67 ± 0.74 | 50.30 ± 1.28 | 49.23 ± 0.43 |
| AV | 50.40 ± 0.65 | 51.48 ± 1.09 | 50.05 ± 0.37 |
| EM | 1981.90 ± 70.02 | 2157.50 ± 117.07 | 1985.76 ± 40.65 |
| EINT | 62.77 ± 2.37 | 65.76 ± 3.78 | 63.13 ± 1.37 |

$^{a,b}$Values with different superscripts within the same row differ significantly ($P < 0.05$)

**Table 4**. Association of the UCP genotypes on breeding values of growth and egg production traits (Mean ± S.E.).

| Traits | Genotype TT | Genotype TC | Genotype CC |
|---|---|---|---|
| WSM | 13.74 ± 4.26 | 22.44 ± 6.86 | −22.81 ± 6.80 |
| ASM | $-21.37 \pm 0.98^{ab}$ | $-19.80 \pm 1.61^{a}$ | $-24.57 \pm 0.61^{b}$ |
| EN | 14.10 ± 0.28 | 14.26 ± 0.45 | 14.43 ± 0.17 |
| EW28 | 0.85 ± 0.2 | 0.96 ± 0.33 | 0.94 ± 0.12 |
| EW30 | 0.58 ± 0.24 | 0.38 ± 0.11 | 0.58 ± 0.15 |
| AV | 1.29 ± 0.17 | 1.50 ± 0.28 | 1.24 ± 0.10 |

$^{a,b}$Values with different superscripts within the same row differ significantly ($P < 0.05$)

## ACKNOWLEDGEMENTS

The authors thank all the team who worked on this study in the experiments and by providing technical assistance in the laboratory.

## REFERENCES

1 Saltzman, E. and Roberts, S.B. (1995) The Role of Energy Expenditure in Energy Regulation: Findings from a Decade of Research. Nutrition Reviews, 53, 209-220.

2 Toubro, S., Sorensen, T.I.A., Ronn, B. and Astrup, A. (1996) Twenty Four-Hour Energy Expenditure: The Role of Body Composition, Thyroid Status, Sympathetic Activity, and Family Membership. Journal of Clinical Endocrinology and Metabolism, 81, 2670-2674.

3 Himms-Hagen, J. (1990) Brown Adipose Tissue Thermogenesis: Interdisciplinary Studies. Fatigue & Fracture of Engineering Materials and Structures, 4, 2890-2898.

4 Adams, S. (2002) Uncoupling Protein Homologs: Emerging Views of Physiological Function. Nutrition, 130, 711-714.

5 Krauss, S., Zhang, C.Y. and Lowell, B.B. (2002) A Significant Portion of Mitochondrial Proton Leak in Intact Thymocytes Depends on Expression of UCP2. Proceedings of the National Academy of Sciences of the United States of America, 99, 118-122.

6 Ricquier, D. and Bouillaud, F. (1997) The Mitochondrial Uncoupling Protein: Structural and Genetic Studies. Proceedings of the National Academy of Sciences of the United States of America-Biological Sciences, 56, 83-108.

7 Liu, Y.J., Liu, P.Y., Long, J., Lu, Y., Elze, L., Recker, R.R. and Deng, H.W. (2005) Linkage and Association Analyses of the UCP 3 Gene with Obesity Phenotypes in Caucasian Families. Physiological Genomics, 22, 197-203.

8 Rudofsky, G.J., Schroedter, A., Schlotterer, A., Voron'ko, O.E., Schlimme, M., Tafel, J., Isermann, B.H., Humpert, P.M., Morcos, M., Bierhaus, A., Nawroth, P.P. and Hamann, A. (2006) Functional Polymorphisms of UCP 2 and UCP 3 Are Associated with a Reduced Prevalence of Diabetic Neuropathy in Patients with Type 1 Diabetes. Diabetes Care, 29, 89-94.

9 Gimeno, R.E., Dembski, M., Gimeno, C.J., et al. (1997) Cloning and Characterization of an Uncoupling Protein Homolog: A Potential Molecular Mediator of Human Thermogenesis. Diabetes, 46, 900-906.

10 Shojaei, M., Mohammad Abadi, M.R., Asadi Fozi, M., Dayani, O., Khezri, A. and Akhondi, M. (2011) Association of Growth Trait and Leptin Gene Polymorphism in Kermani Sheep. Journal of Cell and Molecular Research, 2, 67-73.

11 Askari, N., Mohammadabadi, M.R. and Baghizadeh, A. (2011) ISSR Markers for Assessing DNA Polymorphism and Genetic Characterization of Cattle, Goat and Sheep Populations. Iranian Journal of Biotechnology, 9, 222-229.

12 Zamani, P., Akhondi, M., Mohammadabadi, M.R., et al. (2013) Genetic Variation of Mehraban Sheep Using Two Intersimple Sequence Repeat (ISSR) Markers. African Journal of Biotechnology, 10, 1812-1817.

13 Abadi, M.M., Askari, N., Baghizadeh, A. and Esmailizadeh, A. (2009) A Directed Search around Caprine Candidate Loci Provided Evidence for Microsatellites Linkage to Growth and Cashmere Yield in Rayini Goats. Small Ruminant Research, 81, 146-151.

14 Gilmour, A.R., Gogel, B.J., Cullis, B.R. and Thompson, R. (2006) ASReml User Guide. Release 2.0.

15 Raimbault, S., Dridi, S., Denjean, F., Lachuer, J., Couplan, E., Bouillaud, F., Bordas, A., Duchamp, C., Taouis, M. and Ricquier D. (2001) An Uncoulping Protein Homologue Putatively Involved in Facultative Muscle Thermogenesis in Birds. Biochemical Journal, 353, 441-444.

16 Evock-Clover, C.M., Poch, S.M., Richards, M.P., Ashwell, C.M. and McMurtry, J.P. (2002) Expression of an Uncoupling Protein Gene Homolog in Chickens. Comparative Biochemistry and Physiology a-Molecular and Integrative Physiology, 133, 345-358.

17 Dulloo, A.G. and Samec, S. (2001) Uncoupling Proteins: Their Roles in Adaptive Thermogenesis and Substrate Metabolism Reconsidered. British Journal of Nutrition, 86, 123-139.

18 Himms-Hagen, J. and Harper, M.E. (2001) Physiological Role of UCP3 May Be Export of Fatty Acids from Mitochondria When Fatty Acid Oxidation Predominates: A Hypothesis. Proceedings of the Society for Experimental Biology and Medicine, 226, 78-84.

19 Sherman, E.L., Nkrumah, J.D., Murdoch, B.M., Li, C., Wang, Z., Fu, A. and Moore, S.S. (2008) Polymorphisms and Haplotypes in the Bovine Neuropeptide Y, Growth Hormone Receptor, Ghrelin, Insulin-Like Growth Factor 2, and Uncoupling Proteins 2 and 3 Genes and Their Associations with Measures of Growth, Performance, Feed Efficiency, and Carcass Merit in Beef Cattle. Journal of Animal Science, 86, 1-16.

20 Han, R.H. (2008) Study on the Polymorphisms in UCP3, IGF2, CAPN1 Gene and Its Relationship with Carcass and Meat Quality Traits in Qinchuan Cattle and Its Hybrid Cattle (Dissertation). Northwest A & F University.

21 Li, Q.L. (2006) Study on the Relationship between Polymorphism of UCP3 Gene and Biochemical Index in Serum of Bovine (Dissertation). Northwest A & F University.

22 Boss, O., Hagen, T. and Lowell, B.B. (2000) Uncoupling Proteins 2 and 3, Potential Regulators of Mitochondrial Energy Metabolism. Diabetes, 49, 143-156.

23 Guo, J.J., Liu, Y.J., Li, M.X., Yang, Y.J., Recker, R.R. and Deng, H.W. (2005) Linkage Exclusion Analysis of Two Candidate Regions on Chromosomes 7 and 11: Leptin and UCP2/UCP3 Are Not QTLs for Obesity in US Caucasians. Biochemical and Biophysical Research Communications, 332, 602-608.

24 Argyropoulos, G. and Harper, M.E. (2002) Uncoupling Proteins and Thermoregulation. Journal of Applied Physiology, 92, 2187-2198.

25 Fleury, C., Neverova, M., Collins, S., et al. (1997) Uncoupling Protein-2: A Novel Gene Linked to Obesity and Hyperinsulinemia. Nature Genetics, 15, 269-272.

26 York, B., Truett, A.A., Monteiro, M.P., et al. (1999) Gene-Environment Interaction: A Significant Diet-Dependent Obesity Locus Demonstrated in a Congenic Segment on Mouse Chromosome 7. Mammalian Genome, 10, 457-462.

27 Taylor, B.A. and Phillips, S.J. (1996) Detection of Obesity QTLs on Mouse Chromosome-1 and Chromosome-7 by Selective DNA Pooling. Genomics, 34, 389-398.

28 Vignal, A., Milan, D., San Cristobal, M. and Eggen, A. (2002) A Review on SNP and Other Types of Molecular Markers and Their Use in Animal Genetics. Genetics Selection Evolution, 34, 275-305.

# CHAPTER 11

# Increased genetic gains in sheep, beef and dairy breeding programs from using female reproductive technologies combined with optimal contribution selection and genomic breeding values

**Tom Granleese, Samuel A. Clark, Andrew A. Swan, and Julius H. J. van der Werf**

[1]Sheep Cooperative Research Centre, Armidale 2351, Australia.

[2] School of Environmental and Rural Science, University of New England, Armidale 2351, Australia.

[3] Animal Genetics and Breeding Unit, Armidale 2351, Australia.

## ABSTRACT

### Background

Female reproductive technologies such as multiple ovulation and embryo transfer (MOET) and juvenile in vitro embryo production and embryo transfer (JIVET) can boost rates of genetic gain but they can also increase rates of inbreeding. Inbreeding can be managed using the principles of optimal contribution selection (OCS), which maximizes genetic gain while placing a penalty on the rate of inbreeding. We evaluated the potential benefits and synergies that exist between genomic selection (GS) and reproductive technologies under OCS for sheep and cattle breeding programs.

## Methods

Various breeding program scenarios were simulated stochastically including: (1) a sheep breeding program for the selection of a single trait that could be measured either early or late in life; (2) a beef breeding program with an early or late trait; and (3) a dairy breeding program with a sex limited trait. OCS was applied using a range of penalties (severe to no penalty) on co-ancestry of selection candidates, with the possibility of using multiple ovulation and embryo transfer (MOET) and/or juvenile in vitro embryo production and embryo transfer (JIVET) for females. Each breeding program was simulated with and without genomic selection.

## Results

All breeding programs could be penalized to result in an inbreeding rate of 1 % increase per generation. The addition of MOET to artificial insemination or natural breeding (AI/N), without the use of GS yielded an extra 25 to 60 % genetic gain. The further addition of JIVET did not yield an extra genetic gain. When GS was used, MOET and MOET + JIVET programs increased rates of genetic gain by 38 to 76 % and 51 to 81 % compared to AI/N, respectively.

## Conclusions

Large increases in genetic gain were found across species when female reproductive technologies combined with genomic selection were applied and inbreeding was managed, especially for breeding programs that focus on the selection of traits measured late in life or that are sex-limited. Optimal contribution selection was an effective tool to optimally allocate different combinations of reproductive technologies. Applying a range of penalties to co-ancestry of selection candidates allows a comprehensive exploration of the inbreeding vs. genetic gain space.

# 1. BACKGROUND

Female reproductive technologies such as multiple ovulation and embryo transfer (MOET) and juvenile in vitro fertilization and embryo transfer (JIVET) can be used by breeders to accelerate genetic gain in livestock breeding programs. These technologies allow an increase of selection intensity placed on females and they can also reduce the optimal age at which animals are selected, and thus decrease generation intervals in the breeding program.

Prior to genomic selection (GS) [1], MOET breeding programs were reported to increase genetic gain in dairy breeding programs by up to 30 % by increasing the selection intensity of females [2]. Furthermore, if MOET was used and the age of selection candidates decreased, breeding schemes would increase genetic gain by another 9 % [3]. These gains were similar also in beef and sheep breeding programs, for which MOET was estimated to yield an extra 67 to 100 % of genetic gain for beef [4, 5] and 17 to 74 % for sheep [6–8]. However, all these studies agreed that MOET breeding programs would also increase annual rates of inbreeding by up to 110 % compared to traditional mating programs. Some of these increased rates of inbreeding could have even been underestimated due to reductions in genetic variance and to the assumption that the size of a family and family numbers were constant [9]. For example, the stochastic simulation of Villanueva et al. [10] showed that inbreeding rates increased by 17 % when there is variability between donors in embryo production compared to when embryo numbers are assumed fixed.

When JIVET became commercially viable in the 1990s, selection of donors for JIVET was often based on estimated breeding values (EBV) that had low accuracy and that were highly correlated among siblings due to the great emphasis put on common family information. In this scenario, it is difficult to realize high rates of genetic gain and maintain sustainable rates of inbreeding. Marker-assisted selection [11] and more recently genomic selection [1] have facilitated greater accuracy of EBV and lower correlations of EBV between relatives [12], particularly in younger stock that have not had any performance measurements yet, or if the trait is sex-limited (i.e. milk production), is hard-to-measure (i.e. carcass traits), or has a low heritability (i.e. reproduction).

Recently, some studies have assessed some of the synergies that may exist between the use of reproductive technologies and genomic selection. Using stochastic simulation, Pederson et al. [13] found that genomic selection and MOET on 50 % of nucleus dairy breeding cows resulted in a 23 % increase in genetic gain with no significant increase in inbreeding. Pryce et al. [14] used a deterministic model for a dairy breeding program to show a 98 % increase in genetic gain per year when exclusively using JIVET and genomic selection compared to a progeny test program, but with an increase of inbreeding rate by 65 %. It is important for practical breeding programs to consider numbers of female selection candidates consigned to MOET and JIVET since both are expensive and labor intensive procedures. It is also important to investigate strategies to slow rates of inbreeding which can be accelerated by these technologies.

It is clear from previous studies [3–8, 10, 13, 14] that increasing selection intensity for females in breeding programs can significantly increase rates of inbreeding. Optimal contribution selection (OCS) is a method that balances longer-term genetic merit and genetic diversity in breeding populations. OCS can determine the optimal levels of the genetic contributions of selection candidates but these are often limited by physiological parameters under natural mating. OCS could, however, be an effective tool to assign female reproductive technologies to individual females in breeding programs, while also assigning optimal contributions to males. For a fair comparison and a realistic assessment of the value of reproductive technologies and GS and their potential synergies, comparisons of breeding programs should be done at similar inbreeding rates.

The method developed by Wray and Goddard [15] that applies multiple penalties to co-ancestry while allocating optimal genetic contributions from selection candidates allows the entire selection space between genetic gain and inbreeding to be explored rather than targeting a specific point as demonstrated by Meuwissen [16]. Wray and Goddard's [15] method is useful to investigate the optimal balance between genetic gain and inbreeding. Since OCS allows genetic diversity to be maintained and genetic contributions to be optimized, the synergies that exist between GS and reproductive technologies can be further explored via simulation of breeding programs.

The objective of this study was to assess the value of using female reproduction technologies combined with GS and compare their benefits in beef and dairy cattle and sheep breeding programs, when the aim is to increase rates of genetic gain while keeping inbreeding at sustainable levels. We applied OCS with various penalties on inbreeding and optimized the number of females to be used in MOET or JIVET with regard to genetic gain and inbreeding.

# 2. METHODS

## 2.1 Simulation

Stochastic simulation was used to model a number of closed nucleus breeding schemes. The number of progeny born each year in a closed nucleus was 250, 250 and 600 for sheep, beef and dairy cattle, respectively. For each scenario, we generated a base population of unrelated animals, and subsequently established a 20-year breeding program with overlapping generations. There was an annual random death rate of 10 %. Phenotypes and selection were for a single trait for each breeding program. The genetic values for the base individuals were simulated using a polygenic model as:

$$a_i = z\sigma_a,$$

where $z$ is a random variable drawn from a standard normal distribution and $\sigma_a$ is the genetic standard deviation of the trait. Breeding values for the subsequent generations were obtained using:

$$a_i = ((a_{si} + a_{di})/2) + MS_i,$$

where $a_{si}$ and $a_{di}$ are the true breeding values of the sire and dam of animal $i$, respectively, and $MS_i$ is the Mendelian sampling effect for individual $i$, which was simulated as:

$$MS_i = z \cdot \sigma_a \sqrt{(.5(1 - \overline{F}_i))},$$

where $\overline{F_i}$ is the average inbreeding coefficient of the parents of individual $i$. Phenotypes were then simulated by adding a random error term to $a_i$, such that each trait had a heritability of 0.3 in the base population.

We simulated a number of single-trait scenarios across species that differed in gestation length and whether trait measurement was before or after selection or sex-limited. Phenotypic information was available on selection candidates when the trait could be measured: (1) on both sexes within 6 months of age in sheep (e.g. weight and scan traits); (2) at 2 years of age in sheep on both sexes (e.g. adult wool production); (3) within the first year in beef cattle on both sexes (e.g. weaning and yearling weight); (4) at 19 months of age, after sexual maturity, in beef cattle on both sexes (e.g. 600 day live weight); (5) a sex-limited trait measured on females at 27 months of age (e.g. milk production).

## 2.2 Eligibility and selection of males

### 2.2.1 Sheep

Rams were eligible for selection at 7 months of age and again at 19 months of age. Thereafter, they were culled if not selected at 19 months of age.

### 2.2.2 Beef cattle

Bulls were eligible for selection at 15 and 27 months of age. Thereafter, they were culled if not selected at 27 months of age.

### 2.2.3 Dairy cattle

In breeding programs without GS, bulls were eligible for pre-selection for progeny testing at 15 months of age. Young bulls were selected for progeny testing by conducting the OCS procedure with the 75 highest allocations of contributions selected. By the age of 4.5 years, each bull had 50 daughters (in external herds) with a lactation record. The progeny tested bulls were first eligible for mating in the closed nucleus at 4 years of age and all were kept as selection candidates at 5-years of age. Any bull not selected at 6 years of age or older was culled and all bulls were culled at 9 years of age. Similar to Pryce et al. [14], breeding programs that incorporated GS made bulls eligible for mating at 15 months old. After GS, at 15 months of age, OCS of the top 75 males based on optimal contribution was performed and they were then used in a progeny-testing program with 50 recorded daughters, similar to the breeding program without GS. In breeding programs where GS was used, all bulls that were selected for progeny-testing were eligible for selection every year up to 5 years of age.

### 2.2.4 Genomic selection

Some breeding programs had a scenario where genomic selection was used for all animals born in the nucleus. Genomic information was modeled following the method of Dekkers [17], which simulates a genomic estimated breeding value (GEBV) as a correlated trait with a heritability of 0.99 and a correlation r to the measured trait, where r is the accuracy of the GEBV. The accuracy of the GEBV was assumed to be 0.5 for sheep [18], 0.5 for beef cattle [19] and 0.7 for dairy cattle [20]. All animals received a GEBV at birth and the GEBV was combined with phenotypic information when calculating breeding values.

### 2.2.5 Estimation of breeding values

Each year, EBV were estimated using best linear unbiased prediction (BLUP). The model assumed used was:

$$\mathbf{y} = \mathbf{1}_n\mu + \mathbf{Za} + \mathbf{e},$$

where **y** is a vector of phenotypes, $\mu$ is the mean, $1_n$ is a vector of 1s, **Z** is a design matrix allocating records to breeding values, **a** is a vector of breeding values with $(\mathbf{a}) = \mathbf{A}\sigma_a^2$ where **A** is the numerator relationship matrix and $\sigma_a^2$ is the additive genetic variance, **e** is a random effect. All data from the base population onwards were used for genetic evaluation, including data on the progeny of sires that were progeny-tested.

### 2.2.6 Optimal contribution selection

Optimal contribution selection was implemented using the approach of Wray and Goddard's [15], balancing the average genetic merit (M) of selected individuals with the average co-ancestry among selected individuals (C), both weighted by their contributions to the next generation. Average genetic merit of the selected individuals was evaluated as:

$$\mathbf{M} = \mathbf{x}'\hat{\mathbf{a}},$$

where $\hat{\mathbf{a}}$ is a vector of EBV and **x** is a vector of genetic contributions of selection candidate animals, with values in **x** summing to 0.5 for each of the sexes. The **x** vector contributions for females were capped at a maximum of the equivalent of four offspring per female when MOET or JIVET was used and the equivalent of 1 for the scenarios with only artificial insemination or natural mating (AI/N). After contributions were assigned to females, the number of required matings could be determined for males. The maximum number of matings for a single male was then capped at the total number of female matings. Price and Storn's [21] evolutionary algorithm was used to find optimal solutions for M + C. Once female and male candidates were assigned a mating, they were placed on a mating list. Matings were assigned at random and females were given a chance of a random number of live progeny according to a distribution shown in Table 1. One mating could result in multiple progeny via twins with AI/N, and via embryos implanted in recipients in MOET and JIVET programs.

**Table 1** Probability of producing a certain number of live progeny per female per mating for artificial insemination (AI) or natural mating (AI/N), MOET and JIVET in different species

| Progeny number/program<br>Species | AI/N12[a]<br>Sheep | AI/N[a]<br>Sheep | AI/N[c]<br>Beef | AI[d]<br>Dairy | MOET[b,e]<br>All | JIVET[b]<br>All |
|---|---|---|---|---|---|---|
| 0 | 0.40 | 0.10 | 0.07 | 0.15 | 0.10 | 0.25 |
| 1 | 0.58 | 0.70 | 0.92 | 0.85 | 0.05 | 0.05 |
| 2 | 0.02 | 0.20 | 0.01 | 0.00 | 0.05 | 0.18 |
| 3 | 0.00 | 0.00 | 0.00 | 0.00 | 0.15 | 0.18 |
| 4 | 0.00 | 0.00 | 0.00 | 0.00 | 0.25 | 0.10 |
| 5 | 0.00 | 0.00 | 0.00 | 0.00 | 0.15 | 0.10 |
| 6 | 0.00 | 0.00 | 0.00 | 0.00 | 0.13 | 0.07 |
| 7 | 0.00 | 0.00 | 0.00 | 0.00 | 0.07 | 0.04 |
| 8 | 0.00 | 0.00 | 0.00 | 0.00 | 0.05 | 0.03 |
| Average number of progeny | 0.62 | 1.1 | 0.94 | 0.85 | 4.02 | 8.37[f] |
| Range of progeny possible | 0–2 | 0–2 | 0–2 | 0–1 | 0–8 | 0–24 |

*AI/N7* probability for a ewe at 12 months of age of having a lamb

[a] [22]

[b] [23]

[c] [24]

[d] [25]

[e] [26]

[f] Predicted average of total progeny of three JIVET matings

Inbreeding rates were managed by penalizing the average co-ancestry among selected animals, which was computed as:

$$C = \lambda \mathbf{x}'\mathbf{A}\mathbf{x},$$

where **A** is the (n × n) relationship matrix among all candidates based on pedigree information and λ is a penalty that can be set to result in different rates of inbreeding. Price and Storn's [21] evolutionary algorithm was then used to find optimal contributions as in the vector **x** by maximizing M + C for a given value of λ. These λ values evaluated included (0, −15, −25, −35, −50, −75, −100, −250, −1000, −9999). The values of λ were selected to explore the entire selection space of inbreeding vs genetic gain, which when combined, would form a frontier of possible outcomes. Smaller penalties will result in higher levels of co-ancestry among selected candidates, with potentially higher rates of genetic gain and harsh penalties will result in higher genetic diversity with lower rates of genetic gain.

### 2.2.7 Breeding programs

Three breeding program structures were simulated that differed in the use of reproductive technologies. The first breeding program used only natural mating and/or AI and was simulated as a control. The second breeding program examined the effect of adding MOET to the breeding program, and the third examined the effect of MOET and JIVET combined. The number of offspring that each dam produced under the various reproductive technologies was randomly assigned from a range of offspring based on previous studies [22–26] (Table 1). Females were only offered one chance of being allocated a reproductive technology in each selection round. If no offspring were born, the female was carried as a dry animal to the next round of selection.

### 2.2.8 Breeding program 1: artificial insemination/natural mating (AI/N)

Ewes were eligible for selection at 7 and 19 months of age. Heifers and cows' were eligible for selection at 15 and 27 months of age in the beef and dairy breeding programs. Ewes and cows were culled if not selected at 19 and 27 months of age, respectively.

### 2.2.9 Breeding program 2: AI/N + MOET

Eligibility of females for AI/N or MOET was the same as in the AI/N breeding program because MOET can only be performed on sexually mature females, including young ewes that are just entering puberty [27]. Females that were selected to contribute four offspring based on OCS were assigned to MOET. Females selected to contribute between one and three offspring were assigned to AI/N. Eligibility and culling of females were the same as in the AI/N breeding programs. In all MOET programs, dams were given the chance of having 0 to 8 offspring, with an average of 4 (Table 1). Females were only assigned one male mate for each AI/N or MOET breeding.

### 2.2.10 Breeding program 3: AI/N + MOET + JIVET

The third breeding program added the possibility for juvenile females to be candidates for selection through the use of JIVET. Only juvenile ewes and heifers were eligible for JIVET and they were not eligible for AI/N or MOET in either species. Juvenile females that were selected in OCS to four offspring were assigned JIVET matings. Sexually mature females that were selected to contribute three offspring were assigned to MOET, females selected to contribute less than three offspring were assigned an AI mating, and females that were selected to contribute less than one offspring were not used.

Due to the possibility of using multiple sires in JIVET, three separate matings were assigned to females for JIVET in the same mating period, whereas AI/N and MOET females only received one mating. Selected juvenile females could have between 0 and 8 offspring for each of the three matings, with an average of 2.8 live offspring per mating (Table 1).

Ewes were 1 month old at selection for JIVET. Sheep breeding programs required a mid-year JIVET mating since the timing of JIVET-on-JIVET programs do not fit into annual breeding programs. During the mid-year mating, all ewe selection candidates (including mature ewes) were entered into OCS but only JIVET allocations were executed despite AI and MOET matings being allocated to the mature ewes. As these matings allocated to mature ewes did not yield any progeny, the following round of selection ignored those allocated matings. OCS took into account if ewes had contributed 6 months earlier. In beef and dairy breeding programs, heifers were 3 months old when assigned to JIVET and did not require a mid-year mating like sheep as cattle have a nine month gestation period.

### 2.2.11 Comparison of breeding programs

Each breeding program was run for 20 years and replicated 75 times. Breeding programs were compared based on average annual genetic gains and average annual rates of inbreeding for years 6 to 20, i.e. after the Bulmer equilibrium [28] was reached. Inbreeding was calculated annually as the average inbreeding coefficient of the entire drop of animals born that year using pedigree information. The average annual inbreeding to be displayed in the figures was calculated as the inbreeding increase per year averaged from years 6 to 20 of the simulation. Inbreeding per generation was also calculated due to overlapping generations existing in the simulations. Inbreeding per generation was calculated by matching the average inbreeding coefficient over the last 10 years with the number of generations in $n$ years where:

$$NGens = \left(\frac{n}{L}\right),$$

where $n$ is the number of breeding years and $L$ is the average generation interval over the final 15 years of the simulated breeding program. Generation interval for each individual was calculated as average age of parents at age of birth. The annual generation interval was calculated as the average generation interval of the entire drop born each year.

## 3. RESULTS

### 3.1 Differences between breeding programs without genomic selection

At moderate (1 % per generation) to high (>1.5 % per generation) levels of inbreeding MOET outperformed AI in all cases by 25 to 50 %. At lower inbreeding levels, MOET provided up to 17 to 65 % more genetic gain than AI/N, particularly for late (Figs. 1, 2) and sex-limited traits (Fig. 3). When inbreeding was heavily penalized, there was little genetic gain or inbreeding. In the early measured sheep trait, JIVET yielded 4 to 10 % more genetic gain than MOET (Fig. 4). However, for a late sheep trait and for both beef traits and a sex-limited dairy trait, there was no significant difference in genetic gain between the JIVET and MOET breeding programs (Figs. 1, 2, 3, 5) although some matings were allocated to JIVET (Table 2). As the penalty on inbreeding increased, generation intervals increased and genetic gain decreased (Table 3). It should also be noted that "unpenalized" inbreeding levels in Tables 3 and 4 are of the results of setting λ to zero which are similar to results from truncation selection.

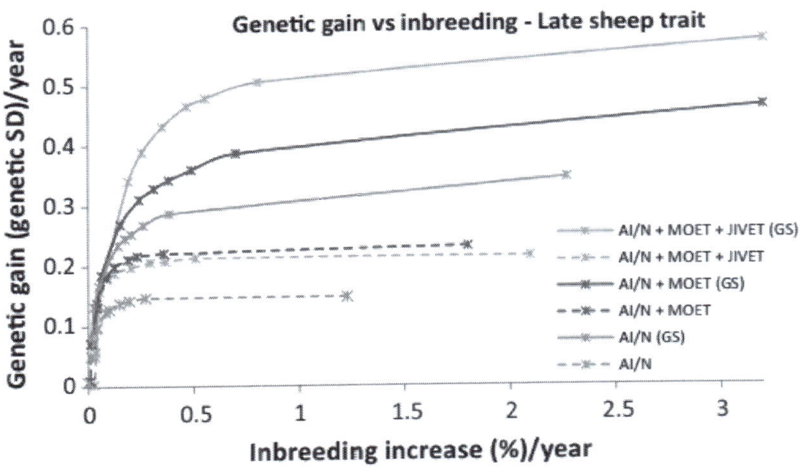


**Figure. 1** Average annual genetic gain (genetic standard deviations) and rate of inbreeding (%) with selection on a late measured trait in sheep using reproductive technologies with and without genomic selection. Data points represent different penalties on coancestry, with no penalty on the far right and a high penalty on the far left

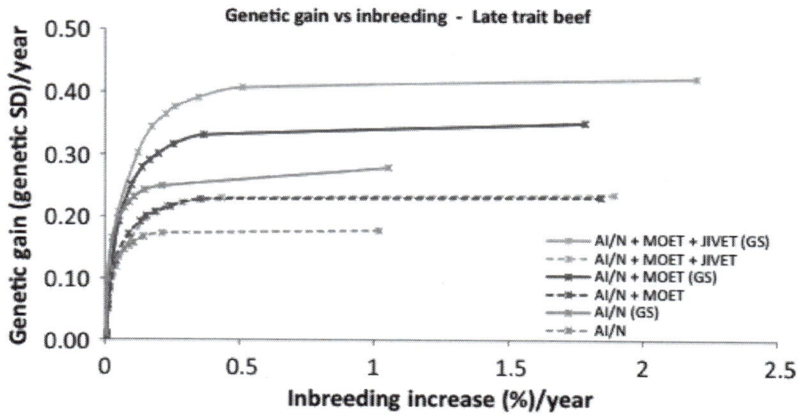


**Figure. 2** Average annual genetic gain (genetic standard deviations) and rate of inbreeding (%) with selection on a late measured trait in beef using reproductive technologies with and without genomic selection. Data points represent different penalties on coancestry, with no penalty on the far right and a high penalty on the far left

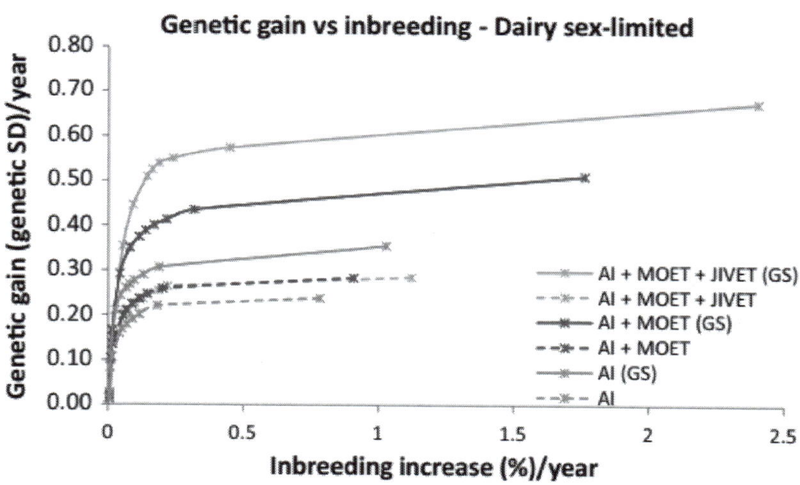


**Figure. 3** Average annual genetic gain (genetic standard deviations) and rate of inbreeding (%) in dairy cattle using reproductive technologies with and without genomic selection. Data points represent different penalties on coancestry, with no penalty on the far right and a high penalty on the far left

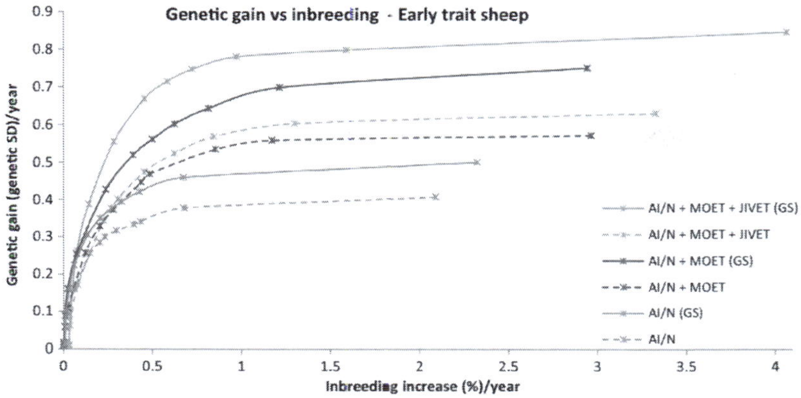


**Figure. 4** Average annual genetic gain (genetic standard deviations) and rate of inbreeding (%) with selection on an early measured trait in sheep using reproductive technologies with and without genomic selection. Data points represent different penalties on coancestry, . with no penalty on the far right and a high penalty on the far left

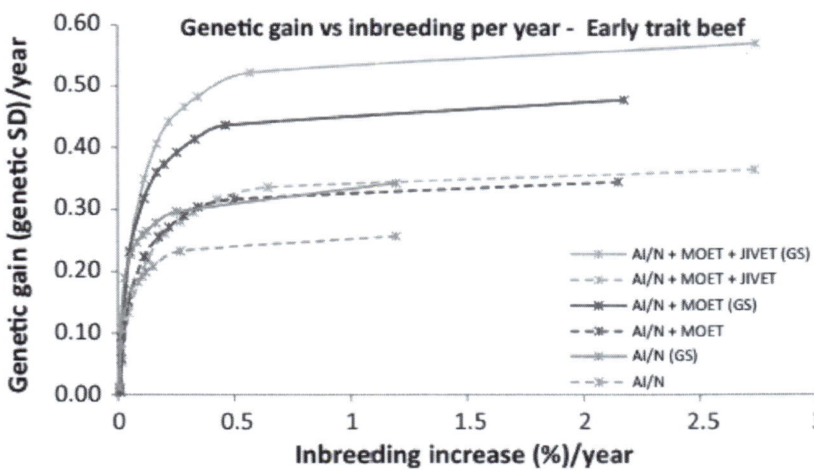


**Figure. 5** Average annual genetic gain (genetic standard deviations) and rate of inbreeding (%) with selection on an early measured trait in beef using reproductive technologies with and without genomic selection. Data points represent different penalties on coancestry, . with no penalty on the far right and a high penalty on the far left

**Table 2** Proportion of selected females assigned each type of mating in each breeding program when rate of inbreeding is 1 % (±0.05 %) per generation for sheep, beef and dairy

| | AI/N | MOET | JIVET | Nb of females selected[a] | Nb of males used[a] | Nb of females per male |
|---|---|---|---|---|---|---|
| Early trait sheep | | | | | | |
| AI/N + MOET + JIVET (GS) | 0.37 | 0.24 | 0.39 | 55 | 23 | 2.4 |
| AI/N + MOET + JIVET | 0.42 | 0.37 | 0.21 | 65 | 24 | 2.7 |
| AI/N + MOET (GS) | 0.41 | 0.59 | | 95 | 23 | 4.0 |
| AI/N + MOET | 0.5 | 0.5 | | 110 | 23 | 4.8 |
| AI/N (GS) | 1 | | | 243 | 28 | 8.7 |
| AI/N | 1 | | | 243 | 31 | 8.0 |
| Late trait sheep | | | | | | |
| AI/N + MOET + JIVET (GS) | 0.36 | 0.18 | 0.46 | 51 | 28 | 1.8 |
| AI/N + MOET + JIVET | 0.42 | 0.29 | 0.29 | 59 | 27 | 2.2 |
| AI/N + MOET (GS) | 0.34 | 0.66 | | 110 | 27 | 4.1 |
| AI/N + MOET | 0.45 | 0.55 | | 110 | 29 | 3.8 |
| AI/N (GS) | 1.00 | | | 245 | 33 | 7.4 |
| AI/N | 1.00 | | | 244 | 32 | 7.7 |
| Early trait beef | | | | | | |
| AI/N + MOET + JIVET (GS) | 0.24 | 0.32 | 0.44 | 47 | 19 | 2.5 |
| AI/N + MOET + JIVET | 0.26 | 0.35 | 0.39 | 57 | 20 | 2.9 |
| AI/N + MOET (GS) | 0.34 | 0.66 | | 98 | 18 | 5.4 |
| AI/N + MOET | 0.34 | 0.66 | | 100 | 19 | 5.3 |
| AI/N (GS) | 1.00 | | | 276 | 14 | 19.7 |
| AI/N | 1.00 | | | 277 | 14 | 19.8 |
| Late trait beef | | | | | | |
| AI/N + MOET + JIVET (GS) | 0.25 | 0.34 | 0.41 | 51 | 14 | 3.6 |
| AI/N + MOET + JIVET | 0.26 | 0.40 | 0.34 | 57 | 15 | 4.1 |
| AI/N + MOET (GS) | 0.32 | 0.68 | | 97 | 15 | 7.7 |
| AI/N + MOET | 0.35 | 0.65 | | 101 | 16 | 5.0 |
| AI/N (GS) | 1.00 | | | 277 | 15 | 18.1 |
| AI/N | 1.00 | | | 279 | 16 | 17.2 |
| Dairy | | | | | | |
| AI + MOET + JIVET (GS) | 0.38 | 0.28 | 0.34 | 143 | 39 | 3.7 |
| AI + MOET + JIVET | 0.47 | 0.35 | 0.18 | 167 | 41 | 4.1 |
| AI + MOET (GS) | 0.47 | 0.53 | | 265 | 40 | 6.7 |
| AI + MOET | 0.44 | 0.56 | | 263 | 34 | 7.7 |
| AI (GS) | 1.00 | | | 646 | 59 | 11.0 |
| AI | 1.00 | | | 652 | 41 | 15.8 |

[a] All standard errors of means (SEM) for total numbers of ewes, cows and males were less than 1.2, 2.7 and 0.18, respectively

## 3.2 Differences between breeding programs with genomic selection

When GS was implemented in breeding programs, programs using JIVET yielded the largest genetic gain, followed by programs using MOET for all species and traits (Table 3; Figs. 1, 2, 3, 4, 5). The largest differences in genetic gain between JIVET and MOET breeding programs were observed for late traits in sheep and beef cattle and for a sex-limited trait in dairy cattle, with differences ranging from 24 to 32 % of genetic gain with MOET at moderate inbreeding levels. Differences in annual genetic gain between MOET and AI/N were largest for early measured traits, ranging from 41 to 43 % of genetic gain with AI (Tables 3, 4). The use of both JIVET and MOET decreased generation intervals, with programs using JIVET displaying the most significant decrease for all species and traits (Table 3).

## 3.3 Impact of genomic selection on breeding programs

The use of GS decreased generation intervals and increased genetic gain across all species and traits (Table 3). Genomic selection increased genetic gain by 37 to 143 % for breeding programs using JIVET, 23 to 77 % for breeding programs using MOET and 23 to 93% for AI/N breeding programs (Table 4). These gains were proportionally higher for late measured traits in sheep (Fig. 1), beef cattle (Fig. 2) and dairy cattle (Fig. 3). The use of GS in early measured traits for sheep (Fig. 4) and beef cattle (Fig. 5) still produced 20 to 40 % more genetic gain when combined with reproductive technologies.

**Table 3** Average generation interval (*L*) and annual genetic gain (ΔG in $\sigma_a$ units) across 75 replicates at various rates of inbreeding (±0.05 %) per generation for sheep, beef and dairy traits

| Inbreeding rate (%/generation) | 0.5 | | 1.0 | | Unpenalized | |
|---|---|---|---|---|---|---|
| | L | ΔG | L | ΔG | L | ΔG |
| Early trait sheep | | | | | | |
| AI/N + MOET + JIVET (GS) | 1.06 | 0.67 | 1.01 | 0.77 | 0.92 | 0.85 |
| AI/N + MOET (GS) | 1.73 | 0.43 | 1.39 | 0.63 | 1.23 | 0.75 |
| AI/N (GS) | 2.04 | 0.39 | 1.94 | 0.44 | 1.84 | 0.50 |
| AI/N + MOET + JIVET | 1.30 | 0.40 | 1.19 | 0.59 | 0.98 | 0.63 |
| AI/N + MOET | 1.73 | 0.37 | 1.48 | 0.51 | 1.27 | 0.57 |
| AI/N | 2.16 | 0.32 | 2.02 | 0.34 | 1.90 | 0.41 |
| Late trait sheep | | | | | | |
| AI/N + MOET + JIVET (GS) | 1.07 | 0.46 | 1.03 | 0.51 | 1.01 | 0.58 |
| AI/N + MOET (GS) | 1.64 | 0.34 | 1.43 | 0.39 | 1.41 | 0.46 |
| AI/N (GS) | 2.25 | 0.25 | 2.08 | 0.29 | 1.96 | 0.35 |
| AI/N + MOET + JIVET | 1.49 | 0.20 | 1.42 | 0.21 | 1.32 | 0.22 |
| AI/N + MOET | 3.20 | 0.21 | 3.12 | 0.22 | 3.03 | 0.23 |
| AI/N | 3.24 | 0.14 | 3.13 | 0.15 | 3.06 | 0.15 |
| Early trait beef | | | | | | |
| AI/N + MOET + JIVET (GS) | 2.28 | 0.41 | 2.1 | 0.49 | 1.99 | 0.57 |
| AI/N + MOET (GS) | 2.93 | 0.36 | 2.71 | 0.41 | 2.59 | 0.48 |
| AI/N (GS) | 3.92 | 0.27 | 3.69 | 0.29 | 3.70 | 0.34 |
| AI/N + MOET + JIVET | 2.63 | 0.25 | 2.27 | 0.31 | 2.19 | 0.36 |

| | | | | | | |
|---|---|---|---|---|---|---|
| AI/N + MOET | 3.36 | 0.21 | 3.19 | 0.23 | 3.00 | 0.23 |
| AI/N | 4.18 | 0.17 | 4.15 | 0.17 | 4.03 | 0.18 |
| Dairy | | | | | | |
| AI + MOET + JIVET (GS) | 1.92 | 0.54 | 1.86 | 0.58 | 1.79 | 0.67 |
| AI + MOET (GS) | 2.66 | 0.40 | 2.58 | 0.44 | 2.46 | 0.51 |
| AI (GS) | 3.55 | 0.28 | 3.41 | 0.32 | 3.35 | 0.35 |
| AI + MOET + JIVET | 4.09 | 0.24 | 3.97 | 0.26 | 3.91 | 0.28 |
| AI + MOET | 4.62 | 0.24 | 4.52 | 0.26 | 4.47 | 0.28 |
| AI | 5.55 | 0.18 | 5.39 | 0.21 | 5.28 | 0.24 |

All standard errors of means (SEM) were less than 0.11 and less than 0.003 for all breeding programs for generation interval and genetic gain, respectively

**Table 4** Increase (%) in rate of genetic gain per year of breeding programs using MOET and JIVET compared to the AI/N breeding program (Repro) and increase (%) in rate of genetic gain per year from the use of genomic selection (GS) on each breeding program at ...

| Inbreeding rate (%/gen) | 0.5 | | 1.0 | | Unpenalized | |
|---|---|---|---|---|---|---|
| | Repro | GS | Repro | GS | Repro | GS |
| Early trait sheep | | | | | | |
| AI/N + MOET + JIVET (GS) | 72 | 68 | 75 | 31 | 70 | 35 |
| AI/N + MOET (GS) | 10 | 16 | 43 | 24 | 50 | 32 |
| AI/N (GS) | | 22 | | 29 | | 22 |
| AI/N + MOET + JIVET | 25 | | 74 | | 54 | |
| AI/N + MOET | 16 | | 50 | | 39 | |
| Late trait sheep | | | | | | |
| AI/N + MOET + JIVET (GS) | 84 | 130 | 76 | 143 | 66 | 163 |
| AI/N + MOET (GS) | 36 | 62 | 34 | 77 | 31 | 100 |
| AI/N (GS) | | 79 | | 93 | | 133 |
| AI/N + MOET + JIVET | 43 | | 40 | | 47 | |
| AI/N + MOET | 50 | | 47 | | 53 | |
| Early trait beef | | | | | | |
| AI/N + MOET + JIVET (GS) | 52 | 64 | 69 | 58 | 68 | 58 |
| AI/N + MOET (GS) | 33 | 38 | 41 | 37 | 41 | 41 |
| AI/N (GS) | | 35 | | 38 | | 31 |
| AI/N + MOET + JIVET | 25 | | 48 | | 38 | |
| AI/N + MOET | 30 | | 43 | | 31 | |
| Late trait beef | | | | | | |
| AI/N + MOET + JIVET (GS) | 64 | 71 | 58 | 78 | 50 | 75 |
| AI/N + MOET (GS) | 32 | 38 | 27 | 43 | 25 | 52 |
| AI/N (GS) | | 29 | | 53 | | 56 |
| AI/N + MOET + JIVET | 24 | | 35 | | 33 | |
| AI/N + MOET | 24 | | 35 | | 28 | |
| Dairy | | | | | | |
| AI + MOET + JIVET (GS) | 93 | 125 | 81 | 123 | 91 | 139 |
| AI + MOET (GS) | 43 | 67 | 38 | 69 | 46 | 82 |
| AI (GS) | | 56 | | 52 | | 46 |
| AI + MOET + JIVET | 33 | | 24 | | 17 | |
| AI + MOET | 33 | | 24 | | 17 | |

When inbreeding was not heavily penalized (>0.5 % inbreeding per generation), the allocation of reproductive technologies increased with the use of GS. When comparing breeding programs with and without GS, there were more JIVET matings with GS at the expense of MOET matings, and this re-allocation was more pronounced for late measured traits and in dairy cattle (Table 3).

# 4. DISCUSSION

## 4.1 Breeding programs

Our results show that, as in prior studies, using female reproductive technologies increases genetic gain when compared to traditional breeding programs. Moreover, we found that the use of female reproductive technologies in combination with the implementation of genomic selection increased rates of genetic gain even more. However, we also observed in all scenarios that if inbreeding was not penalized, large inbreeding coefficients could be accumulated over time. This study demonstrates the benefits of optimizing the allocation of females to reproductive technologies through the use of optimal contribution selection, which resulted in increased rates of genetic gain under sustainable rates of inbreeding. Combining MOET with GS yielded 25 to 50 % more genetic gain compared to AI and combining JIVET with GS increased genetic gain up to 143 % without increasing the rate of inbreeding.

Genomic selection resulted in clear benefits between breeding strategies for all traits, species and breeding programs using different technologies. When comparing breeding programs with and without GS, increases in genetic gains from the use of GS ranged from 33 to 143 % for JIVET programs, from 37 to 77 % for MOET programs and from 29 to 93 % for AI/N programs (Tables 3, 4). When comparing programs that used GS, JIVET led to 15 to 43 % greater genetic gain than MOET breeding programs. We also observed a 38 to 76 % greater genetic gain for MOET breeding programs compared to AI/N breeding programs when GS was implemented in both breeding programs. These gains of MOET over AI and JIVET over MOET were due to greater accuracies of EBV at selection of young selection candidates and lower correlations between the EBV of siblings. Genomic selection allows an increase of selection accuracy at the age of first selection and this increase is greatest for late and sex-limited traits. Genomic information increases accuracy of EBV by explaining within-family variance due to Mendelian sampling [12], which is important when selecting animals early because without GS there is little information about this component. Moreover, the Bulmer effect reduces the between-family variance and, therefore, contributes less to future genetic gains than the within-family variance [29]. Dekkers [17] concluded that highly accurate (0.8) GEBV could increase genetic gain by 21 and 81 % over a traditional phenotypic selection program for traits with a heritability of 0.3 and 0.1, respectively. These extra gains from GS depend on whether traits have been measured on selection candidates at the time of first selection. In dairy cattle, we used accurate GEBV for GS (0.7) and observed that GS increased rates of genetic gain in traditional AI programs by 52 % (Tables 3, 4). Using GS in sheep and beef cattle programs, with a lower accuracy of GEBV (0.5), increased rates of genetic gain by 29 to 93 %, depending on species and when the trait was measured (Tables 3, 4). Genomic predictions in beef cattle and sheep are typically less accurate than in dairy cattle due to the extensive and multi-breed aspect of these industries, which makes it more difficult to generate training populations that sustain high accuracies. Studies in sheep that assumed maximum accuracies of GEBV of 0.5 showed that GS increased genetic gain by 21 % in terminal sire breeds and up to 39 % in fine wool Merino sheep [30].

## 4.2 Optimal contribution selection

While many previous studies have evaluated the short-term benefits of reproductive technologies, they did not all explore how the rate of inbreeding can be minimized or reduced. Wray and Goddard [15] developed a selection principle to manage rates of inbreeding through a penalty method. Other studies implement OCS by maximizing genetic gain at a specific rate of inbreeding [16, 31, 32]. Together, these methods have demonstrated that maintaining a high level of genetic gain at set rates of inbreeding can be achieved. We used a range of inbreeding penalties to create 'frontiers' of genetic gain versus inbreeding, thereby clearly showing the balance that can be achieved (Figs. 1, 2, 3, 4, 5), rather than targeting specific inbreeding rates. We were then able to choose specific inbreeding rates per generation to compare scenarios. While genetic gain somewhat plateaued when the rate of inbreeding increased to 1 % per generation (Table 3), managers of breeding programs must decide which point on the frontier suits their breeding objective.

Our study is unique in that it uses OCS to optimize genetic gain with inbreeding in livestock breeding programs that use female reproductive technologies. It also optimally assigned reproductive technologies according to how much each mating contributes to the objective function of maximizing gain while penalizing inbreeding. Table 2 demonstrates the proportion of females that were allocated to different reproductive technologies. We observed that, with the added accuracy that is provided by GS, the proportions of MOET allocations did not change in AI + MOET breeding programs. However we observed that with the use of GS in AI + MOET + JIVET breeding programs increased JIVET allocations by up to 86, 20 and 88 % and decreased MOET allocations by up to 38, 15 and 20 % for sheep, beef cattle and dairy cattle breeding programs, respectively. In conjunction with a shift in the selection of more JIVET candidates, we observed a decrease in generation intervals (Table 3). We also observed that the number of males selected each year differed only slightly between programs (Table 2). It should be noted that males are generally used below their maximum reproductive capacity. The use of males below their natural or artificial insemination capabilities was needed to maintain diversity without sacrificing too much genetic gain. The concept of reducing inbreeding while implementing reproductive technologies by using more sires under their reproductive capacity has been described previously by Kinghorn [33], Arendonk and Bijma [34], Pryce et al. [14] and Lillehammer et al. [35] in dairy, and by Brash et al. [8] in sheep breeding schemes.

## 4.3 Comparing with previous studies

Our study estimated that the benefit of MOET in sheep breeding programs without using GS, was up to 50 % while maintaining the rate of inbreeding at 1 % per generation. Previous studies [7, 8] using MOET in sheep breeding programs also predicted such gains but without restricting inbreeding rates. Our beef MOET breeding programs achieved 35 to 43 % more genetic gain than AI programs without GS which was less than previous studies, which reported extra gains of 67 to 138 %

[7, 9, 10], but we achieved these gains at the same rate of inbreeding per generation, while rates of inbreeding increased by 87 to 300 % in the aforementioned studies. Use of MOET in our dairy breeding program resulted in 24 % extra genetic gain without increasing the rate of inbreeding. This is similar to the values reported by Kinghorn [33] and Leitch et al. [3], who estimated similar gains in dairy cattle nucleus breeding programs that implemented MOET compared to traditional AI nucleus breeding programs but with larger increases in rates of inbreeding (1.7 to 4.2 % increase per year). Without GS, all traits and species benefited from the use of MOET. However, except for the early measured trait in sheep, which yielded 4 to 10 % greater annual genetic gain, the use of JIVET without GS did not yield extra genetic gain compared to AI + MOET programs in late trait sheep, beef and dairy cattle breeding programs, although some JIVET matings were allocated to juveniles (Table 2). The lower allocation of JIVET matings is likely due to the low accuracy of EBV and the high correlation between EBV of full siblings at a young age. This demonstrates that, even without accounting for the additional cost of JIVET, this technology may not give an advantage to breeding programs when using optimal contribution selection across age classes, at least without GS.

Previous studies have also evaluated breeding programs that implement female reproductive technologies with GS in comparison to traditional programs. Pryce et al. [14] deterministically modeled a closed nucleus of Holstein cattle in which 300 juvenile females that were submitted to JIVET were mated only to genomically selected sires and compared the rates of genetic gain and inbreeding to a progeny-testing AI breeding program. The JIVET program resulted in a 131 % increase in genetic gain and a 7 % decrease in rate of inbreeding, which they obtained by increasing the number of sires of bulls and cows. Their breeding program resulted in 0.59 genetic standard deviations of genetic gain per year at a 0.17 % rate of inbreeding per year. The most comparable scenario in our study is the AI + MOET + JIVET (GS) program vs. AI. With a rate of inbreeding of 0.17 % per year for both programs, we observed that AI + MOET + JIVET (GS) resulted in a 145 % greater genetic gain per year than the AI breeding program, with gains of 0.54 and 0.22 genetic standard deviations per year, respectively (Fig. 5). The genetic gain that we obtained in the dairy cattle AI + MOET + JIVET (GS) breeding program was not quite as large as that in Pryce et al. [14]. This could be due to our stochastic simulation also using MOET and AI technologies which resulted in higher generation intervals and lower selection intensities compared to Pryce et al.'s [14] 100 % JIVET each year.

## 4.4 Rates of inbreeding

Breeding programs resulting in high rates of inbreeding should be avoided due to the likelihood of declines in fitness [36, 37] and homozygous recessive genotypes that are lethal [38]. However, if inbreeding is penalized too heavily, genetic diversity is maintained but with low rates of genetic gain (Figs. 1, 2, 3, 4, 5). Therefore, a compromise must be made to achieve desirable amounts of genetic gain without increasing inbreeding too much. Figures 1, 2, 3, 4 and 5 show rates of inbreeding per year, but past studies suggest using inbreeding rate per generation as a metric as it can be used to compare different breeding programs with different generation intervals. Bijma [39] suggested restricting rates of inbreeding to 1 % per generation, which was supported by Buch et al. [40].This desired rate of inbreeding will differ between producers due to varying concerns about the importance of inbreeding, the size of the nucleus, the ability to import unrelated animals into the nucleus, the effective population size of the breed as a whole, and the incidence of recessive homozygous genetic diseases [41].

## 4.5 Simulation restrictions

In this study, GEBV were simulated as a correlated trait with a heritability of 1 and a genetic correlation with the trait equal to the accuracy of the GEBV [24]. Genomic selection can also be simulated by simulating quantitative trait loci (QTL) and genetic markers. However, because we did not simulate genotypes, predicting the linkage disequilibrium (LD) patterns across generations [42] between markers and QTL was not addressed in our simulations. Family sizes that increase in a nucleus over generations through the use of MOET and JIVET could slightly increase the accuracy of the individual GEBV of those family members [43]. However, predicting the changes in gene frequencies of the nucleus population using an infinitesimal model could be difficult to determine over a period longer than three generations [44]. An alternative method that simulates the actual marker genotypes could be used to capture the change in LD structure over time due to selection [45]. Some studies [46–48] have studied the efficacy of genomic selection when performed over many generations, and where only a few QTL control genetic variation. They suggest that the effects of markers do not change but the proportion of genetic variance explained by them will decline, which means that the effects of genomic selection may be over-estimated in our simulations. However, it has been suggested that hundreds or thousands of QTL may control variation for a given trait [49, 50]. Heffner et al. [51] demonstrated that if all QTL effects are small, as in an infinitesimal model, genomic selection could maintain accuracies over multiple generations due to the change in LD between QTL and markers being less important over a longer time period. Therefore, given both sides of the argument it is difficult to predict whether the method of Dekkers [17] over- or under-estimates genetic gains from genomic selection.

In our study, we used a pedigree-derived relationships matrix (**A**) to measure inbreeding and also during OCS. Some authors have shown that using a genomic relationship matrix (**G**) in OCS has some advantages over using **A** by achieving similar rates of genetic gain with slightly lower rates of inbreeding [52, 53]. This is achieved because **G** accounts for differences in the level of relatedness within families. Using **G** would however increase computing time for an already time-consuming simulation.

This study did not take the costs of technology into account. If they were, we expect that the use of reproductive technologies would decrease. In some cases, reproductive technologies were assigned although the additional genetic benefit was limited, in particular when inbreeding was highly penalized. Therefore, this study evaluated the maximum use and benefit in terms of genetic gain. Including costs will, however, give a more realistic picture of potential use. A cost benefit assessment will also focus more on how genetic gain is translated into actual monetary benefit for the breeder who carries the burden of the additional investment.

# 5. CONCLUSIONS

Reproductive technologies combined with genomic selection can substantially enhance rates of genetic gain without compromising rates of inbreeding when optimal contribution selection is used with an incurred penalty on future co-ancestry. Without genomic selection, the use of JIVET in conjunction with MOET is not beneficial. These results suggest that synergies exist between the applications of reproductive technologies, genomic selection and optimal contribution selection. Optimal contribution selection was an effective tool in optimizing the allocations of combined reproductive technologies. The full exploration of the frontier between inbreeding vs. genetic gain production was facilitated by applying a range of penalties to co-ancestry of selection candidates.

# AUTHORS' CONTRIBUTIONS

TG wrote and carried out the simulations, carried out statistical analysis and drafted the manuscript. SC, AS assisted in computer coding and participated in the design of the study. JvdW conceived the idea of this study, participated in the design of the study and assisted with computer coding. All authors read and approved the final manuscript.

# ACKNOWLEDGEMENTS

The authors acknowledge the University of New England and Cooperative Research Centre for Sheep Industry Innovation for funding.

# COMPLIANCE WITH ETHICAL GUIDELINES

**Competing interests** The authors declare that they have no competing interests.

## REFERENCES

1 Meuwissen THE, Hayes BJ, Goddard ME. Prediction of total genetic value using genome-wide dense marker maps. Genetics. 2001;157:1819–1829.

2 Nicholas FW, Smith C. Increased rates of genetic change in dairy cattle by embryo transfer and splitting.Anim Sci. 1983;36:341–353.

3 Leitch HW, Smith C, Burnside EB, Quinton M. Genetic response and inbreeding with different selection methods and mating designs for nucleus breeding programs of dairy cattle. J Dairy Sci. 1994;77:1702–1718.

4 Gearheart WW, Smith C, Teepker G. Multiple ovulation and embryo manipulation in the improvement of beef cattle: relative theoretical rates of genetic change. J Anim Sci. 1989;67:2863–2871.

5 Land RB, Hill WG. The possible use of superovulation and embryo transfer in cattle to increase response to selection. Anim Sci. 1975;21:1–12.

6 Smith C. Use of embryo transfer in genetic improvement of sheep. Anim Sci. 1986;42:81–86.

7 Horton B. A method of using a genetic algorithm to examine the optimum structure of the Australian sheep breeding industry: open-nucleus breeding systems MOET and AI. Aust J Exp Agr. 1996;36:249–258.

8 Brash LD, Wray NR, Goddard ME. Use of MOET in Merino breeding programmes: a practical and economic appraisal. Anim Sci. 1996;62:241–254.

9 Wray NR, Simm G. The use of embryo transfer to accelerate genetic improvement in beef cattle. In: Proceedings of the 4th World Congress on Genetics Applied to Livestock Production, 23–27 July 1990: Edinburgh; 1990, p. 315–8.

10 Villanueva B, Woolliams JA, Simm G. The effect of improved reproductive performance on genetic gain and inbreeding in MOET breeding schemes for beef cattle. Genet Sel Evol. 1995;27:347–363.

11 Meuwissen THE, Goddard ME. The use of marker haplotypes in animal breeding schemes. Genet Sel Evol. 1996;28:161–176.

12 Clark SA, Kinghorn BP, Hickey JM, van der Werf JHJ. The effect of genomic information on optimal contribution selection in livestock breeding programs. Genet Sel Evol. 2013;45:44.

13 Pedersen LD, Kargo M, Berg P, Voergaard J, Buch LH, Sørensen AC. Genomic selection strategies in dairy cattle breeding programmes: sexed semen cannot replace multiple ovulation and embryo transfer as superior reproductive technology. J Anim Breed Genet. 2012;129:152–163.

14 Pryce JE, Goddard ME, Raadsma HW, Hayes BJ. Deterministic models of breeding scheme designs that incorporate genomic selection. J Dairy Sci. 2010;93:5455–5466.

15 Wray NR, Goddard ME. Increasing long-term response to selection. Genet Sel Evol. 1994;26:431–451.

16 Meuwissen THE. Maximizing the response of selection with a predefined rate of inbreeding. J Anim Sci.1997;75:934–940.

17 Dekkers JCM. Prediction of response to marker-assisted and genomic selection using selection index theory. J Anim Breed Genet. 2007;124:331–341.

18 Swan AA, Brown DJ, Daetwyler HD, Hayes BJ, Kelly M, Moghaddar N, et al. Genomic evaluations in the Australian sheep industry. In: Proceedings of the 10th World Congress on Genetics Applied to Livestock Production: 17–22 August 2014; Vancouver; 2014, p. 334.

19 Bolormaa S, Pryce JE, Kemper K, Savin K, Hayes BJ, Barendse W, et al. Accuracy of prediction of genomic breeding values for residual feed intake and carcass and meat quality traits in Bos taurus, Bos indicus, and composite beef cattle. J Anim Sci. 2013;91:3088–3104.

20 Hayes BJ, Bowman PJ, Chamberlain AJ, Verbyla K, Goddard ME. Accuracy of genomic breeding values in multi-breed dairy cattle populations. Genet Sel Evol. 2009;41:51.

21 Price K, Storn R. Differential evolution. Dr Dobb's J. 2004;78:18–24.

22 Bunter KL, Brown DJ. Yearling and adult expressions of reproduction in maternal sheep breeds are genetically different traits. In Proceedings of the twentieth Association for Advancement in Animal Breeding and Genetics: 20–23 October 2013: Napier; 2013, p. 82–85.

23 Armstrong DT, Kotaras PJ, Earl C. Advances in production of embryos in vitro from juvenile and prepubertal oocytes from the calf and lamb. Reprod Fertil Dev. 1997;9:333–339.

24 Houghton PL, Lemenager P, Horstman LA, Hendrix KS, Moss GE. Effects of body composition, pre-and postpartum energy level and early weaning on reproductive performance of beef cows and preweaning calf gain. J Anim Sci. 1990;68:1438–1446.

25 Buckley F, O'Sullivan K, Mee JF, Evans RD, Dillon P. Relationships among milk yield, body condition, cow weight, and reproduction in spring-calved Holstein-Friesians. J Dairy Sci. 2003;86:2308–2319.

26 Hasler JF. Factors affecting frozen and fresh embryo transfer pregnancy rates in cattle. Theriogenology.2001;56:1401–1415.

27 Nicholas FW. Genetic improvement through reproductive technology. Anim Reprod Sci. 1996;42:205–214.

28 Bulmer MG. The effect of selection on genetic variability. Am Nat. 1971;105:201–211.

29 Bijma P. Accuracies of estimated breeding values from ordinary genetic evaluations do not reflect the correlation between true and estimated breeding values in selected populations. J Anim Breed Genet.2012;129:345–358.

30 van der Werf JHJ. Potential benefit of genomic selection in sheep. In: Proceedings of the 18th Conference of the Association for the Advancement of Animal Breeding and Genetics: 28 September–1 October 2009: Barossa Valley; 2009, p. 38–41.

31 Meuwissen THE, Sonesson AK. Maximizing the response of selection with a predefined rate of inbreeding: overlapping generations. J Anim Sci. 1998;76:2575–2583.

32 Grundy B, Villanueva B, Woolliams JA. Dynamic selection for maximizing response with constrained inbreeding in schemes with overlapping generations. Anim Sci. 2000;70:373–382.

33 Kinghorn BP, Smith C, Dekkers JCM. Potential genetic gains in dairy cattle with gamete harvesting and in vitro fertilization. J Dairy Sci. 1991;74:611–622.

34 van Arendonk JAM, Bijma P. Factors affecting commercial application of embryo technologies in dairy cattle in Europe—a modelling approach. Theriogenology. 2003;59:635–649.

35 Lillehammer M, Meuwissen THE, Sonesson AK. A comparison of dairy cattle breeding designs that use genomic selection. J Dairy Sci. 2011;94:493–500.

36 Smith LA, Cassell BG, Pearson RE. The effects of inbreeding on the lifetime performance of dairy cattle.J Dairy Sci. 1998;81:2729–2737.

37 Van Raden PM, Olson KM, Null DJ, Hutchison JL. Harmful recessive effects on fertility detected by absence of homozygous haplotypes. J Dairy Sci. 2011;94:6153–6161.

38 Shariflou MR, Wade CM, Windsor PA, Tammen I, James JW, Nicholas FW. Lethal genetic disorder in Poll Merino/Merino sheep in Australia. Aust Vet J. 2011;89:254–259.

39 Bijma P. Long-term genetic contributions. PhD Thesis, Wageningen University, Department of Animal Sciences; 2000.

40 Buch LH, Sørensen MK, Berg P, Pedersen LD, Sørensen AC. Genomic selection strategies in dairy cattle: strong positive interaction between use of genotypic information and intensive use of young bulls on genetic gain. J Anim Breed Genet. 2012;129:138–151.

41 Goddard ME. Optimal effective population size for the global population of Black and White dairy cattle.J Dairy Sci. 1992;75:2902–2911.

42 Farnir F, Coppieters W, Arranz JJ, Berzi P, Cambisano N, Grisart B, et al. Extensive genome-wide linkage disequilibrium in cattle. Genome Res. 2000;10:220–227.

43 Risch N, Teng J. The relative power of family-based and case-control designs for linkage disequilibrium studies of complex human diseases I. DNA pooling. Genome Res. 1998;8:1273–1288.

44 Meuwissen THE. Optimisation of dairy cattle breeding plans with increased female reproductive rates. PhD Thesis, Wageningen University, Research Institute for Animal Production; 1990.

45 VanRaden PM. Efficient methods to compute genomic predictions. J Dairy Sci. 2008;91:4414–4423.

46 Muir WM. Comparison of genomic and traditional BLUP-estimated breeding value accuracy and selection response under alternative trait and genomic parameters. J Anim Breed Genet. 2007;124:342–355.

47 Sonesson AK, Meuwissen THE. Testing strategies for genomic selection in aquaculture breeding programs. Genet Sel Evol. 2009;41:37.

48 Zhang XS, Hill WG. Predictions of patterns of response to artificial selection in lines derived from natural populations. Genetics. 2005;169:411–425.

49 Hayes BJ, Chamberlain AJ, McPartlan H, Macleod I, Sethuraman L, Goddard ME. Accuracy of marker-assisted selection with single markers and marker haplotypes in cattle. Genet Res. 2007;89:215–220.

50 Goddard ME. Genomic selection: prediction of accuracy and maximisation of long term response.Genetica. 2009;136:245–257.

51 Heffner EL, Jannink JL, Sorrells ME. Genomic selection accuracy using multifamily prediction models in a wheat breeding program. Plant Genome. 2011;4:65–75.

52 Dagnachew BS, Meuwissen THE. An iterative algorithm for optimum contribution selection in large scale breeding programs. In: Proceedings of the 10th World Congress of Genetics Applied to Livestock Production: 15–22 August 2014; Vancouver; 2014, p. 023.

53 Nielsen HM, Sonesson AK, Meuwissen THE. Optimum contribution selection using traditional best linear unbiased prediction and genomic breeding values in aquaculture breeding schemes. J Anim Sci.2011;89:630–638.

# CHAPTER 11

# Accuracies of genomically estimated breeding values from pure-breed and across-breed predictions in Australian beef cattle

**Vinzent Boerner, David J Johnston, and Bruce Tier**

Animal Genetics and Breeding Unit, University of New England, Armidale 2351, NSW, Australia

## ABSTRACT

### Background

The major obstacles for the implementation of genomic selection in Australian beef cattle are the variety of breeds and in general, small numbers of genotyped and phenotyped individuals per breed. The Australian Beef Cooperative Research Center (Beef CRC) investigated these issues by deriving genomic prediction equations (PE) from a training set of animals that covers a range of breeds and crosses including Angus, Murray Grey, Shorthorn, Hereford, Brahman, Belmont Red, Santa Gertrudis and Tropical Composite. This paper presents accuracies of genomically estimated breeding values (GEBV) that were calculated from these PE in the commercial pure-breed beef cattle seed stock sector.

## Methods

PE derived by the Beef CRC from multi-breed and pure-breed training populations were applied to genotyped Angus, Limousin and Brahman sires and young animals, but with no pure-breed Limousin in the training population. The accuracy of the resulting GEBV was assessed by their genetic correlation to their phenotypic target trait in a bi-variate REML approach that models GEBV as trait observations.

## Results

Accuracies of most GEBV for Angus and Brahman were between 0.1 and 0.4, with accuracies for abattoir carcass traits generally greater than for live animal body composition traits and reproduction traits. Estimated accuracies greater than 0.5 were only observed for Brahman abattoir carcass traits and for Angus carcass rib fat. Averaged across traits within breeds, accuracies of GEBV were highest when PE from the pooled across-breed training population were used. However, for the Angus and Brahman breeds the difference in accuracy from using pure-breed PE was small. For the Limousin breed no reasonable results could be achieved for any trait.

## Conclusion

Although accuracies were generally low compared to published accuracies estimated within breeds, they are in line with those derived in other multi-breed populations. Thus PE developed by the Beef CRC can contribute to the implementation of genomic selection in Australian beef cattle breeding.

# 1. BACKGROUND

Genomic selection (GS) has been introduced into breeding schemes of many livestock species [1-3]. The advantages of GS compared to conventional breeding schemes can be summarised as: (i) shortening the generation interval because genomically estimated breeding values (GEBV) can be calculated early in life, (ii) estimation of GEBV for all genotyped individuals of a species/breed for difficult to measure traits given a prediction equation (PE) that has been derived from a related population and (iii) increased accuracy of estimated breeding values for lowly heritable traits [2,4]. Since in beef cattle breeding, selection candidates usually have some of their own performance records before the selection decision is made, the generation interval is usually not a constraint for the genetic progress. Thus, advantages (ii) and (iii) are the key improvements for Australian beef cattle breeding schemes expected from the implementation of GS [5]. In the dairy industry, several conditions have facilitated the implementation of GS: (1) a large number of phenotypes is collected routinely; (2) wide-spread use of artificial insemination facilitates the use of highly accurate conventionally estimated sire breeding values as pseudo-phenotypes; and (3) large breeding organisations can bear the initial cost of genotyping. Currently, these conditions are not met in the Australian beef cattle industry. On the contrary, the Australian beef industry is made up of a large number of breeds and crosses (both *Bos taurus* and *Bos indicus*), breeding organisations are rather small, and records of economically important traits on live animals or carcasses are usually expensive to measure and limited in number, as are the genotypes of phenotyped individuals [5]. This situation is reflected in low numbers of genotyped individuals which are generally not sufficient to calculate accurate within-breed GEBV.

A possible approach to make GS feasible for breeds with a small number of genotypes and phenotypes is the derivation of PE that allow prediction of GEBV across breeds [4,6]. This derivation is usually done on a mixed breed training population that contains individuals of all targeted breeds. Thus, the number of genotyped individuals in the reference population might exceed the total number of genotyped individuals in any single breed, which may allow for a higher power to detect single nucleotide polymorphisms (SNPs) in strong linkage disequilibrium (LD) with trait coding quantitative trait loci (QTL). However, whether all or only the small breeds gain from this approach depends on the proportions of each breed in the training population [4]. The across-breed-prediction approach was followed by the Australian Beef Cooperative Research Center (www.beefcrc.com, Beef CRC), which derived PE on a pooled training population of genotyped individuals from eight different cattle breeds and on different cross-breed and pure-breed subsets of this pooled set, in which the genotyped individuals originated from Australian populations of Angus, Murray Grey, Shorthorn, Hereford, Brahman, Belmont Red, Santa Gertrudis, Tropical Composite breeds, and F1 crosses of Brahman with Limousin, Charolais, Angus, Shorthorn and Hereford breeds [7].

Beef CRC PE were derived for the commercial animal breeding sector. The value of the PE for breeders depends on the accuracy of the resulting GEBV. This accuracy is the proportion of the additive genetic variance of the focused phenotypic trait, estimated in the commercial seed stock population, that is explained by the GEBV. Common approaches to assess the accuracy and the PE consist in subdividing of the training data and a subsequent n-fold cross-validation, or the derivation of the PE in generation $i$ and the derivation of the accuracy in generation $i+n, n \geq 1$ [7-11]. However, in both cases, the accuracy is usually calculated as a product moment correlation, sometimes scaled by some value. Whether accuracies obtained in this way are also achievable in the commercial seed stock population depends on a variety of factors such as the genetic distance between the commercial and the training population [12] and the sample size of the training population. Another approach to obtain an estimator of the proportion of the additive genetic variance in the seed stock population explained by the GEBV is to apply PE to genotyped seed stock individuals, model the resulting GEBV as trait observations in a bi-variate approach together with their phenotypic target trait and assess the co-variances by restricted maximum likelihood (REML) or Gibbs sampling [13-16]. This approach accounts for various sources of bias in parameter estimation, including genetic trends, relatedness between individuals, inbreeding and differences in accuracy of EBV. In addition, the genetic correlation between the phenotypic target trait and the GEBV that is obtained this way is an indispensable part of the blending of EBV with GEBV.

The aim of this work was to determine whether Beef CRC PE derived within and across breeds facilitate the introduction of GS in the Australian commercial beef cattle seed stock herds. For this purpose, GEBV were calculated for genotyped seed stock animals of Australian Angus, Limousin and Brahman breeds, thus, subsets of those populations Beef CRC PE have been derived for, and their accuracy was assessed as the genetic correlation to their phenotypic target trait in a bi-variate REML approach.

# 2. METHODS

## 2.1 Genomically estimated breeding values

### 2.1.1 Prediction equations

The assembly of the training population, genotyping of training individuals and PE derivation were not part of this project. For a detailed description of the PE derivation and the size, breed composition and animal characteristics of the training population see [7].

However, in short, PE supplied to the authors were derived within the Beef CRC on 800K Illumina HD Bovine genotypes in a 5-fold cross-validation genomic best linear unbiased prediction (GBLUP) approach with phenotypic records as response variables [7]. PE were developed for the following traits: post-weaning live weight (g.WW), live weight on feedlot entry (g.YW), live weight on feedlot exit (g.FW), carcass rib fat (g.CRIB), carcass P8 fat (g.CP8), carcass intra-muscular fat (g.CIMF) and carcass weight (g.CWT). For a list of GEBV and their abbreviations see Table 1. Genotyped training animals originated from Australian populations of the Angus, Murray Grey, Shorthorn, Hereford, Brahman, Belmont Red, Santa Gertrudis, Tropical Composite breeds, and F1 crosses of the Brahman breed with the Limousin, Charolais, Angus, Shorthorn and Hereford breeds, and included cows, steers and bulls. For each of the above traits, PE were derived on four sets of individuals: all genotyped animals across breeds (ALL), Angus only (ANGUS), *Bos taurus* only (Angus, Murray Grey, Hereford, Shorthorn) (BOSTAURUS) and Brahman only (BRAHMAN).

**Table 1** GEBV and phenotypic traits and the used trait abbreviations

| Trait | Abbreviation |
|---|---|
| GEBV post-weaning live weight | g.WW |
| GEBV live weight at feedlot period start | g.YW |
| GEBV live weight at feedlot period end | g.FW |
| GEBV scan eye muscle area | g.SEMA |
| GEBV scan rib fat | g.SRIB |
| GEBV scan P8 fat | g.SP8 |
| GEBV carcass rib fat | g.CRIB |
| GEBV carcass p8 fat | g.CP8 |
| GEBV carcass intramuscular fat | g.CIMF |
| GEBV carcass weight | g.CWT |
| Phenotypic 200 day weight | p.WW |
| Phenotypic 400 day weight | p.YW |
| Phenotypic 600 day weight | p.FW |
| Phenotypic bull eye muscle area | p.BEMA |
| Phenotypic heifer/steer eye muscle area | p.HEMA |
| Phenotypic bull rib fat | p.BRIB |
| Phenotypic heifer/steer rib fat | p.HRIB |
| Phenotypic bull P8 fat | p.BP8 |
| Phenotypic heifer/steer P8 fat | p.HP8 |
| Phenotypic carcass rib fat | p.CRIB |
| Phenotypic carcass P8 fat | p.CP8 |
| Phenotypic carcass intra-muscular fat percentage | p.CIMF |
| Phenotypic carcass weight | p.CWT |

## 2.2 Genotypes and genomically estimated breeding values of commercial seed stock animals

The ALL, ANGUS, BOSTAURUS and BRAHMAN PE were applied to genotypes of commercial seed stock animals that originated from

Australian populations of the Angus, Limousin and Brahman breeds. None of these genotyped individuals were in the training population. This set of animals will be referred to as "validation set". For all three breeds, the validation set consisted of widely used sires and animals from the current generation. The numbers of individuals (sires/animals in the current generation) in each breed sample were 1582 (383/1199) for Angus, 782 (368/414) for Limousin, and 400 (108/302) for Brahman. After removing individuals that did not match the breed specific pedigrees, the validation sets consisted of 1487 Angus, 721 Limousin and 400 Brahman individuals. Genotypes of all validation animals were obtained using the Illumina 50K Bead Chip. To apply Beef CRC PE, all genotypes were imputed from 50K to 800K. Imputation was done with a population-based approach [17] using 800K Beef CRC genotypes [7], and 2500 800K genotypes of Limousin, Charolais and Simmental individuals, supplied by the Irish Cattle Breeding Federation (ICBF), as reference genotypes. The population approach was necessary because many of animals in the current generation were not registered at the time of imputation, which made it impossible to exploit possible duo or trio structures in the data because of unknown parents. Finally, GEBV were calculated by applying the above described PE to the animals' genotypes.

## 2.3 Compilation of phenotypic datasets

The phenotypic datasets and pedigree data of all three breeds were obtained from databases of their respective breed societies: Angus Australia, Australian Limousin Breeders' Society and Australian Brahman Breeders' Association. All phenotypic data were adjusted for systematic effects as described in [18]. For all traits and breeds, the number of records in the phenotypic datasets exceeded those of the GEBV datasets. For Angus, some individuals that were used in the training population were also part of the phenotypic datasets (see Table 2).

Phenotypic traits included in the analysis were 200-day weight (p.WW), 400-day weight (p.YW), 600-day weight (p.FW), bull's scan eye muscle area (p.BEMA), heifer's scan eye muscle area (p.HEMA), bull's scan rib fat (p.BRIB), heifer's scan rib fat (p.HRIB), bull's scan P8 fat (p.BP8), heifer's scan P8 fat (p.HP8), carcass rib fat (p.CRIB), carcass P8 fat (p.CP8), carcass intra-muscular fat (p.CIMF) and carcass weight (p.CWT). Note that not all traits were available for each breed. For a list of phenotypic traits and their abbreviations see Table 1.

**Table 2** Parameters of phenotypic traits

| Trait[1] | Angus | | | | Limousin | | | Brahman | | |
|---|---|---|---|---|---|---|---|---|---|---|
| | N[2] | n[3] | $\bar{x}^4$ | $s^5$ | N | $\bar{x}$ | s | N | $\bar{x}$ | s |
| p.WW | 120928 | 73 | 252.3 | 38.6 | 83210 | 239.4 | 43.6 | 145558 | 210.4 | 39.7 |
| p.YW | 81428 | 39 | 383.6 | 71.5 | 60044 | 380.0 | 79.9 | 67115 | 280.9 | 84.7 |
| p.FW | 114170 | 54 | 535.2 | 103.5 | 38545 | 521.8 | 115.7 | 70955 | 393.6 | 111.8 |
| p.BEMA | 88265 | 0 | 83.0 | 10.8 | 3450 | 101.8 | 14.4 | 6655 | 74.6 | 19.5 |
| p.HEMA | 101221 | 250 | 62.9 | 9.9 | 3237 | 75.2 | 14.1 | 4494 | 53.6 | 11.7 |
| p.BRIB | 88256 | 0 | 3.7 | 1.5 | 3381 | 2.9 | 1.3 | 6097 | 3.8 | 2.5 |
| p.HRIB | 101360 | 289 | 5.3 | 2.3 | 3229 | 3.4 | 1.8 | 4018 | 3.4 | 2.0 |
| p.BP8 | 88087 | 0 | 4.7 | 2.0 | 3443 | 3.4 | 1.7 | 6292 | 5.3 | 3.4 |
| p.HP8 | 101530 | 289 | 6.8 | 3.2 | 3254 | 4.4 | 2.6 | 4184 | 5.3 | 3.2 |
| p.CRIB | 1203 | 573 | 12.3 | 5.5 | - | - | - | 1486 | 7.7 | 3.7 |
| p.CP8 | 2183 | 0 | 21.0 | 6.4 | 1057 | 11.2 | 4.5 | 1575 | 14.0 | 5.1 |
| p.CIMF | 2822 | 0 | 7.2 | 2.6 | 986 | 3.4 | 1.5 | 1584 | 2.8 | 1.3 |
| p.CWT | 3839 | 0 | 400.4 | 47.4 | 1082 | 303.2 | 67.7 | - | - | - |

[1] for trait abbreviations see Table 1, [2] number of phenotypic observations, [3] number of Angus individuals with a phenotypic record in this data set which have been used for deriving ANGUS, BOSTAURUS and ALL prediction equation, [4] mean, [5] standard deviation. Note that no Limousin individuals were used in the training set and all phenotyped Brahman individuals which where part of the training set were excluded from the analysis.

Contemporary groups were formed as defined in [18] but all single record groups were deleted. For p.WW, p.YW and p.FW, records were excluded if the sire, dam, maternal grandsire or embryo transfer recipient dam was unknown. Multiple records of these traits were also deleted except for the first. In order to decrease the computational demand, the number of Angus records for p.WW and p.YW was further reduced as follows: records were kept only if the recorded individual was part of the validation set, or was a direct progeny of an individual in the validation set, or was in a contemporary group with an individual that belonged to one of the latter two groups.

## 2.4 Estimation of the variance components

Variances and variance ratios were obtained from bi-variate REML analysis for which each phenotypic trait was analysed in conjunction with its assigned GEBV. The fitted model for p.WW, p.YW and p.FW and their respective GEBV was:

$$\begin{pmatrix} \mathrm{y} \\ \mathrm{y_{GEBV}} \end{pmatrix} = \begin{pmatrix} \mathrm{X} & 0 \\ 0 & 1 \end{pmatrix} \begin{pmatrix} \mathrm{b_p} \\ \mathrm{b_g} \end{pmatrix} + \begin{pmatrix} \mathrm{Z_d} & 0 & \mathrm{Z_m} & \mathrm{Z_p} \\ 0 & \mathrm{Z_g} & 0 & 0 \end{pmatrix} \begin{pmatrix} \mathrm{u_d} \\ \mathrm{u_g} \\ \mathrm{u_m} \\ \mathrm{p} \end{pmatrix} + \begin{pmatrix} \mathrm{e_p} \\ \mathrm{e_g} \end{pmatrix} \tag{1}$$

where **y**, $\mathbf{y_{GEBV}}$, $\mathbf{b_p}$, $\mathbf{b_g}$, $\mathbf{u_g}$, $\mathbf{u_d}$, $\mathbf{u_m}$, **p**, $\mathbf{e_p}$ and $\mathbf{e_g}$ are vectors of phenotypic observations, GEBV, fixed effects of the phenotypic trait, the mean of the GEBV, random direct additive genetic effects of the GEBV, random direct additive genetic, random maternal additive genetic, random maternal environmental and random residual effects of the phenotypic trait and random residual effects of the GEBV, respectively. **X**, $\mathbf{Z_d}$,$\mathbf{Z_g}$, $\mathbf{Z_m}$ and $\mathbf{Z_p}$ are incidence matrices relating the effects to their phenotypic observations or GEBV, respectively, and **1** is a vector of 1s. Random effects in the model were assumed to be multivariate normally distributed with

$$\begin{pmatrix} \mathrm{u_d} \\ \mathrm{u_g} \\ \mathrm{u_m} \\ \mathrm{p} \\ \mathrm{e_p} \\ \mathrm{e_g} \end{pmatrix} \sim \mathrm{N}\left( \begin{bmatrix} 0 \\ 0 \\ 0 \\ 0 \\ 0 \\ 0 \end{bmatrix}, \begin{bmatrix} \mathbf{A}\sigma^2_{\mathrm{a}} & \mathbf{A}\sigma_{\mathrm{a,g}} & \mathbf{A}\sigma_{\mathrm{a,m}} & 0 & 0 & 0 \\ \mathbf{A}\sigma_{\mathrm{a,g}} & \mathbf{A}\sigma^2_{\mathrm{g}} & \mathbf{A}\sigma_{\mathrm{g,m}} & 0 & 0 & 0 \\ \mathbf{A}\sigma_{\mathrm{a,m}} & \mathbf{A}\sigma_{\mathrm{g,m}} & \mathbf{A}\sigma^2_{\mathrm{m}} & 0 & 0 & 0 \\ 0 & 0 & 0 & \mathbf{I}\sigma^2_{p} & 0 & 0 \\ 0 & 0 & 0 & 0 & \mathbf{I}\sigma^2_{\mathrm{e_p}} & \mathbf{I}\sigma_{\mathrm{e_{p,g}}} \\ 0 & 0 & 0 & 0 & \mathbf{I}\sigma_{\mathrm{e_{p,g}}} & \mathbf{I}\sigma^2_{\mathrm{e_g}} \end{bmatrix} \right), \tag{2}$$

where **A** is the numerator relationship matrix constructed such that every individual with a phenotypic or GEBV observation had at least three generations of ancestors in the pedigree if available, and **I** is an identity matrix. $\sigma^2_{\mathrm{a}}$ is the variance of the direct additive genetic effect of the phenotypic trait, $\sigma^2_{\mathrm{g}}$ is the variance of the direct additive genetic effect of the GEBV, $\sigma^2_{\mathrm{m}}$ is the variance of the maternal additive genetic effect of the phenotypic trait, $\sigma_{\mathrm{a,m}}$ is the covariance between the direct additive genetic effect of the phenotypic trait and the maternal additive genetic effect of the phenotypic trait, $\sigma_{\mathrm{a,g}}$ is the covariance between the direct additive genetic effect of the phenotypic trait and the direct additive genetic effect of the GEBV, $\sigma_{g,m}$ is the covariance between the direct additive genetic effect of the GEBV and the maternal additive genetic effect of the phenotypic trait, $\sigma^2_{p}$ is the variance of the maternal permanent environmental effect, $\sigma^2_{e_p}$ is the variance of the residual effect of the phenotypic trait, $\sigma^2_{e_g}$ is the variance of the residual effect of the GEBV, and $\sigma_{e_{p,g}}$ is the co-variance of the residual effect of the GEBV and the residual effect of the phenotypic trait.

The fitted bi-variate model for all other phenotypic traits and their respective GEBV was:

$$\begin{pmatrix} y \\ y_{GEBV} \end{pmatrix} = \begin{pmatrix} X & 0 \\ 0 & 1 \end{pmatrix}\begin{pmatrix} b_p \\ b_g \end{pmatrix} + \begin{pmatrix} Z_d & 0 \\ 0 & Z_g \end{pmatrix}\begin{pmatrix} u_d \\ u_g \end{pmatrix} + \begin{pmatrix} e_p \\ e_g \end{pmatrix}, \tag{3}$$

with random effects assumed to be multivariate normally distributed with

$$\begin{pmatrix} u_d \\ u_g \\ e_p \\ e_g \end{pmatrix} \sim N\left(\begin{bmatrix} 0 \\ 0 \\ 0 \\ 0 \end{bmatrix}, \begin{bmatrix} A\sigma^2_a & A\sigma_{a,g} & 0 & 0 \\ A\sigma_{a,g} & A\sigma^2_g & 0 & 0 \\ 0 & 0 & I\sigma^2_{e_p} & I\sigma_{e_{p,g}} \\ 0 & 0 & I\sigma_{e_{p,g}} & I\sigma^2_{e_p} \end{bmatrix}\right). \tag{4}$$

For phenotypic traits, contemporary group was the only fixed effect, and for GEBV only the mean was fitted as a fixed effect. Note that the residual covariance was fitted only for the combinations of GEBV and phenotypic traits where a subset of the individuals had observations on both, the GEBV and the phenotypic trait.

## 2.5 Software

Imputation was done with Beagle [17] without exploiting any parent-offspring pair/parent-offspring trio structure. The number of iterations in Beagle was set to 30. Pre- and post-analysis data manipulation was done with R [19] and Sweave [20]. REML analyses were carried out with WOMBAT [21].

# 3. RESULTS

## 3.1 Raw data

Table 2 summarises the results for all phenotypic traits across breeds for the following parameters: number of observations, mean, standard deviation and number of animals that are in common between the phenotypic dataset and the training population. The number of observations for growth traits exceeded those for the difficult and expensive-to-measure carcass traits for all three breeds. Large numbers of records for live animal ultrasound scan traits were available for the Angus breed only, whereas for the Brahman and Limousin breeds records of these traits were almost as limited as those for carcass traits. An overlap between the phenotypic dataset and the training dataset was found for the Angus breed only, which means that in this case the datasets were not totally independent. However, only for p.CRIB, could the proportion of training individuals used in the phenotypic dataset (573 of 1203 phenotypic observations) have caused an upward biased accuracy. For all other traits, this proportion was zero or negligible due to more phenotypic records and less training individuals in the phenotypic dataset. As mentioned above, none of the genotyped validation animals were used in the training population. However, the mean, minimum and maximum relationships between the validation individuals and those training individuals of the same breed, based on a pedigree constructed for these animals three generations back, were equal to 0.014, 0.0 and 0.57 for Angus, and 0.008, 0.0 and 0.57 for Brahman. Note that the mean relationship includes only training individuals of the same breed as the target population. Thus, for mixed-breed training populations, this number is even smaller when all training individuals are included, and if the training population and the target population represent different breeds, all three parameters are equal to 0.

## 3.2 Heritabilities of genomically estimated breeding values

Table 3 summarises heritabilities ($h^2$) and their standard errors for all GEBV. Across traits and PE high $h^2$ of almost 1 with the lowest standard errors were found consistently for the Angus breed only. For the Brahman breed, in most cases the $h^2$ values were below 0.9. Across traits, the lowest $h^2$ values were always estimated for the GEBV calculated from the ANGUS PE, followed by those from the BOSTAURUS PE and BRAHMAN PE. The highest $h^2$ were almost exclusively estimated for ALL PE GEBV, except for g.WW which was below 0.9. In most cases, the standard errors of $h^2$ for the Brahman breed GEBV were above 0.1, and therefore about five times as large as those for Angus, which reflects the size of the Brahman sample. The lowest $h^2$ across traits and PE were found for the Limousin breed with most values below 0.6 and the lowest estimates equal to 0.42 for g.SRIB from ALL PE. In contrast to Brahman, no generally superior or inferior PE could be identified for Limousin. Heritabilities from the uni-variate analysis were not different to those from the bi-variate analysis (results not shown).

## 3.3 Genetic correlations between GEBV and phenotypic traits

Table 4 summarises the genetic correlations ($r_g$) between GEBV and phenotypic traits for the Australian Angus breed. The highest $r_g$ (0.53) was found for p.CRIB :g.CRIB derived from BOSTAURUS PE, the lowest (-0.01) for p.BP8 :g.SP8 derived from BRAHMAN PE, but most values were below 0.2. Across all traits, ALL PE and BOSTAURUS PE yielded the highest $r_g$ followed by ANGUS and BRAHMAN PE, where the ALL PE results almost mirrored those from ANGUS and BOSTAURUS PE. BRAHMAN PE was inferior for carcass traits, whereas for growth traits (except p.FW :g.FW) differences between $r_g$ of

GEBV from different PE were small. As a result of the number of phenotypic observations, standard errors of $r_g$ for growth traits were below 0.1, but much larger for carcass traits for which fewer data were available.

**Table 3** Heritabilities (upper) and standard errors (lower) of GEBV

| Trait[1] | Angus ($n^2$ = 1487) | | | | Limousin (n = 721) | | | | Brahman (n = 400) | | | |
|---|---|---|---|---|---|---|---|---|---|---|---|---|
| | ALL[3] | An[4] | Bt[5] | Br[6] | ALL | An | Bt | Br | ALL | An | Bt | Br |
| g.WW | 0.95 | 1.00 | 0.98 | 0.99 | 0.61 | 0.50 | 0.59 | 0.65 | 0.71 | 0.64 | 0.71 | 0.84 |
| g.YW | 0.96 | 0.99 | 0.96 | 0.99 | 0.56 | 0.46 | 0.42 | 0.70 | 0.99 | 0.68 | 0.74 | 0.96 |
| g.FW | 0.96 | 0.99 | 0.98 | 1.00 | 0.43 | 0.58 | 0.57 | 0.58 | 0.97 | 0.77 | 0.86 | 0.87 |
| g.SEMA | 0.97 | 0.98 | 0.99 | 0.97 | 0.76 | 0.78 | 0.47 | 0.59 | 0.92 | 0.88 | 0.86 | 0.89 |
| g.SRIB | 1.00 | 1.00 | 1.00 | 1.00 | 0.42 | 0.49 | 0.46 | 0.58 | 1.00 | 0.85 | 0.99 | 0.99 |
| g.SP8 | 0.97 | 0.97 | 0.97 | 0.96 | 0.56 | 0.57 | 0.62 | 0.58 | 0.97 | 0.78 | 0.88 | 0.87 |
| g.CRIB | 0.98 | 1.00 | 0.99 | 1.00 | - | - | - | - | 0.96 | 0.96 | 1.00 | 0.99 |
| g.CP8 | 0.97 | 0.99 | 0.96 | 1.00 | 0.46 | 0.66 | 0.54 | 0.66 | 0.96 | 0.94 | 0.81 | 0.87 |
| g.CIMF | 0.99 | 1.00 | 1.00 | 1.00 | 0.65 | 0.54 | 0.76 | 0.59 | 0.90 | 0.83 | 0.92 | 0.92 |
| g.CWT | 0.95 | 1.00 | 0.93 | 1.00 | 0.47 | 0.50 | 0.56 | 0.60 | - | - | - | - |
| g.WW | 0.03 | 0.02 | 0.03 | 0.03 | 0.10 | 0.10 | 0.10 | 0.09 | 0.15 | 0.14 | 0.12 | 0.13 |
| g.YW | 0.03 | 0.03 | 0.03 | 0.03 | 0.10 | 0.10 | 0.10 | 0.09 | 0.11 | 0.14 | 0.13 | 0.11 |
| g.FW | 0.03 | 0.02 | 0.03 | 0.03 | 0.11 | 0.10 | 0.10 | 0.09 | 0.11 | 0.13 | 0.12 | 0.12 |
| g.SEMA | 0.03 | 0.03 | 0.03 | 0.03 | 0.07 | 0.07 | 0.09 | 0.09 | 0.11 | 0.13 | 0.12 | 0.12 |
| g.SRIB | 0.03 | 0.03 | 0.03 | 0.03 | 0.10 | 0.10 | 0.11 | 0.09 | 0.08 | 0.13 | 0.11 | 0.09 |
| g.SP8 | 0.03 | 0.03 | 0.03 | 0.04 | 0.10 | 0.10 | 0.10 | 0.10 | 0.10 | 0.14 | 0.13 | 0.12 |
| g.CRIB | 0.03 | 0.03 | 0.03 | 0.03 | - | - | - | - | 0.10 | 0.10 | 0.10 | 0.09 |
| g.CP8 | 0.03 | 0.03 | 0.03 | 0.03 | 0.10 | 0.09 | 0.10 | 0.08 | 0.09 | 0.11 | 0.13 | 0.09 |
| g.CIMF | 0.02 | 0.02 | 0.02 | 0.03 | 0.09 | 0.09 | 0.08 | 0.09 | 0.10 | 0.13 | 0.11 | 0.11 |
| g.CWT | 0.04 | 0.02 | 0.04 | 0.03 | 0.10 | 0.10 | 0.10 | 0.09 | - | - | - | - |

[1]for trait abbreviations see Table 1, [2]number of individuals with a GEBV, [3]prediction equation derived on a mixed-breed training population, [4]prediction equation derived on a training population of Angus individuals, [5]prediction equation derived on a training population of *Bos taurus* individuals, [6]prediction equation derived on a training population of Brahman individuals. - : for Limousin cattle no date were available.

For the Limousin breed, $r_g$ variec more than for the Angus breed, and their standard errors were much greater (see Table 5). Across traits and PE $r_g$ varied from 0.63 to -0.69 for p.CP8 :g.CP8 estimated from BOSTAURUS PE and BRAHMAN PE, respectively. No clear pattern regarding superior or inferior PE could be identified because $r_g$ varied considerably within traits across PE. For example, $r_g$ of p.WW :g.WW was -0.02 from ALL PE, 0.22 from ANGUS PE, 0.08 from BOSTAURUS PE and -0.03 from BRAHMAN PE.

Table 6 summarises the GEBV $r_g$ for Australian Brahman, which varied across traits and PE from 0.7 (p.CRIB :g.CRIB from ALL PE) to -0.5 (p.CIMF :g.CIMF from ANGUS PE). However, negative $r_g$ were exclusively found in ANGUS and BOSTAURUS PE. Moreover, results from ALL PE almost mirrored those from BRAHMAN PE, whereas

ANGUS and BOSTAURUS PE yielded much smaller or even negative $r_g$. Standard errors decreased with the availability of more phenotypic data (low for carcass traits and high for growth traits), and were similar across PE, except for most carcass traits, for which standard errors from ANGUS and BOSTAURUS PE were double those from ALL and BRAHMAN PE.

**Table 4** Correlation| *standard error* between the direct additive genetic component of the phenotypic trait and GEBV from different prediction equations for Australian Angus Cattle

| Phenotypic[1] | GEBV[1] | ALL[2] | ANG[3] | BT[4] | BRA[5] |
|---|---|---|---|---|---|
| p.WW | g.WW | 0.09\|*0.06* | 0.07\|*0.05* | 0.10\|*0.06* | 0.11\|*0.06* |
| p.YW | g.YW | 0.08\|*0.06* | 0.09\|*0.06* | 0.09\|*0.06* | 0.14\|*0.06* |
| p.FW | g.FW | 0.19\|*0.05* | 0.18\|*0.05* | 0.21\|*0.05* | 0.11\|*0.06* |
| p.BEMA | g.SEMA | 0.16\|*0.06* | 0.18\|*0.06* | 0.16\|*0.06* | 0.01\|*0.06* |
| p.HEMA | g.SEMA | 0.15\|*0.05* | 0.10\|*0.05* | 0.13\|*0.05* | 0.09\|*0.06* |
| p.BRIB | g.SRIB | 0.26\|*0.06* | 0.25\|*0.06* | 0.25\|*0.05* | 0.09\|*0.06* |
| p.HRIB | g.SRIB | 0.20\|*0.05* | 0.19\|*0.05* | 0.20\|*0.05* | 0.05\|*0.05* |
| p.BP8 | g.SP8 | 0.25\|*0.06* | 0.27\|*0.06* | 0.25\|*0.06* | -0.01\|*0.06* |
| p.HP8 | g.SP8 | 0.21\|*0.05* | 0.23\|*0.05* | 0.21\|*0.05* | 0.02\|*0.05* |
| p.CRIB | g.CRIB | 0.51\|*0.21* | 0.36\|*0.17* | 0.53\|*0.21* | 0.12\|*0.35* |
| p.CP8 | g.CP8 | 0.36\|*0.20* | 0.27\|*0.17* | 0.34\|*0.19* | 0.12\|*0.23* |
| p.CIMF | g.CIMF | 0.33\|*0.12* | 0.29\|*0.12* | 0.36\|*0.13* | 0.00\|*0.17* |
| p.CWT | g.CWT | 0.25\|*0.15* | 0.30\|*0.12* | 0.25\|*0.14* | -0.00\|*0.17* |

[1]for trait abbreviations see Table 1, [2]prediction equation derived on a mixed-breed training population, [3]prediction equation derived on a training population of Angus individuals, [4]prediction equation derived on a training population of *Bos taurus* individuals, [5]prediction equation derived on a training population of Brahman individuals.

**Table 5** Correlation| *standard error* between the direct additive genetic component of the phenotypic trait and GEBV from different prediction equations for Australian Limousin cattle

| Phenotypic[1] | GEBV[1] | ALL[2] | ANG[3] | BT[4] | BRA[5] |
|---|---|---|---|---|---|
| p.WW | g.WW | -0.01\|*0.09* | 0.22\|*0.10* | 0.08\|*0.10* | -0.03\|*0.09* |
| p.YW | g.YW | -0.15\|*0.10* | 0.02\|*0.11* | 0.06\|*0.11* | -0.11\|*0.09* |
| p.FW | g.FW | -0.03\|*0.11* | -0.01\|*0.10* | -0.03\|*0.10* | 0.06\|*0.10* |
| p.BEMA | g.SEMA | 0.28\|*0.14* | -0.29\|*0.13* | -0.01\|*0.19* | 0.01\|*0.16* |
| p.HEMA | g.SEMA | -0.29\|*0.16* | -0.59\|*0.10* | -0.25\|*0.17* | -0.25\|*0.16* |
| p.BRIB | g.SRIB | -0.40\|*0.19* | -0.14\|*0.17* | -0.36\|*0.17* | -0.23\|*0.16* |
| p.HRIB | g.SRIB | 0.40\|*0.13* | -0.38\|*0.11* | -0.06\|*0.17* | -0.42\|*0.11* |
| p.BP8 | g.SP8 | -0.13\|*0.16* | -0.27\|*0.12* | -0.24\|*0.13* | -0.20\|*0.15* |
| p.HP8 | g.SP8 | 0.08\|*0.16* | 0.01\|*0.15* | 0.02\|*0.15* | -0.05\|*0.15* |
| p.CP8 | g.CP8 | 0.08\|*0.36* | -0.14\|*0.29* | 0.63\|*0.28* | -0.69\|*0.18* |
| p.CIMF | g.CIMF | 0.47\|*0.25* | -0.26\|*0.30* | -0.04\|*0.24* | 0.23\|*0.30* |
| p.CWT | g.CWT | 0.24\|*0.29* | 0.06\|*0.31* | 0.34\|*0.28* | 0.17\|*0.29* |

[1]for trait abbreviations see Table 1, [2]prediction equation derived on a mixed-breed training population, [3]prediction equation derived on a training population of Angus individuals, [4]prediction equation derived in a training population of *Bos taurus* individuals, [5]prediction equation derived on a training population of Brahman individuals.

# 4. DISCUSSION

## 4.1 Genetic correlations

Across-breed PE were derived by the Beef CRC to facilitate the implementation of GS in the Australian beef cattle industry, which is made difficult by the large number of breeds, small numbers of individuals with genotypes and/or phenotypes per breed and their unequal distribution across breeds, and the widespread use of cross-breeds. It has been proposed that PE derived from large mixed-breed samples may circumvent these problems. Moreover, the power of detection of SNPs in high LD with a QTL that affects the phenotype of interest is expected to increase when mixed-breed data is used [8,22].

**Table 6** Correlation| *standard error* between the direct additive genetic component of the phenotypic trait and GEBV from different prediction equations for Australian Brahman cattle

| Phenotypic[1] | GEBV[1] | ALL[2] | ANG[3] | BT[4] | BRA[5] |
|---|---|---|---|---|---|
| p.WW | g.WW | 0.27\|*0.10* | 0.07\|*0.11* | 0.15\|*0.10* | 0.23\|*0.09* |
| p.YW | g.YW | 0.19\|*0.10* | 0.14\|*0.11* | 0.14\|*0.11* | 0.20\|*0.09* |
| p.FW | g.FW | 0.20\|*0.09* | -0.17\|*0.11* | -0.07\|*0.10* | 0.20\|*0.10* |
| p.BEMA | g.SEMA | -0.08\|*0.22* | 0.19\|*0.26* | -0.24\|*0.26* | -0.17\|*0.23* |
| p.HEMA | g.SEMA | -0.04\|*0.24* | -0.23\|*0.26* | -0.10\|*0.27* | -0.06\|*0.26* |
| p.BRIB | g.SRIB | 0.45\|*0.17* | -0.08\|*0.23* | 0.01\|*0.22* | 0.41\|*0.17* |
| p.HRIB | g.SRIB | 0.18\|*0.22* | -0.29\|*0.26* | -0.14\|*0.25* | 0.19\|*0.23* |
| p.BP8 | g.SP8 | 0.34\|*0.20* | -0.08\|*0.21* | -0.16\|*0.21* | 0.24\|*0.20* |
| p.HP8 | g.SP8 | 0.32\|*0.21* | 0.20\|*0.23* | -0.09\|*0.23* | 0.30\|*0.21* |
| p.CRIB | g.CRIB | 0.70\|*0.20* | 0.10\|*0.46* | 0.44\|*0.43* | 0.65\|*0.21* |
| p.CP8 | g.CP8 | 0.57\|*0.19* | 0.46\|*0.30* | 0.44\|*0.34* | 0.34\|*0.24* |
| p.CIMF | g.CIMF | 0.56\|*0.27* | -0.50\|*0.37* | -0.20\|*0.41* | 0.36\|*0.25* |

[1]for trait abbreviations see Table 1, [2]prediction equation derived on a mixed-breed training population, [3]prediction equation derived an a training population of Angus individuals, [4]prediction equation derived on a training population of *Bos taurus* individuals, [5]prediction equation derived on a training population of Brahman individuals.

Accuracy of GEBV derived from Beef CRC PE in a 5-fold cross-validation approach were published by [7]. Since Beef CRC prediction equations were developed for application in the Australian commercial beef cattle seed stock herds, the aim of this work was to validate accuracies in these herds via a bi-variate REML approach. In addition, estimated parameters are a precondition for blending estimated breeding values with GEBV. Accuracies of GEBV from ALL PE for Australian Angus were calculated as REML genetic correlations between GEBV and their phenotypic target traits, and were found to be considerably different to those given by [7]. For instance cross-validation accuracies of g.WW, g.YW, g.SRIB and g.SP8 for the Angus breed were reported to be equal to 0.27, 0.42, 0.42 and 0.5 respectively, while the values estimated in our study were 0.09, 0.08, 0.26 and 0.25 respectively. On the contrary, cross-validation accuracies for g.SEMA, g.CIMF and g.CWT reported by [7] were equal to those found here (0.15 vs. 0.15, 0.31 vs. 0.33) or lower (0.16 vs. 0.25). However, the standard errors of our results do not allow us to draw an unambiguous conclusion on whether the latter three estimates are significantly different from 0. For the Brahman breed, accuracies published by [7] for g.WW, g.YW and g.SEMA were also considerably higher than those obtained from ALL PE. In contrast, for g.SP8 and g.CIMF, ALL PE yielded higher accuracies (0.34 vs. 0.19, 0.56 vs. 0.27). However, for these GEBV the standard errors do not support the conclusion that accuracies are significantly different from 0. One possible reason for the differences is the genetic distance between our validation dataset and the training dataset. The Beef CRC collection of genotypes started in the early 2000. Thus, the distance between some validation and training genotypes might represent several generations. Moreover, some genotypes were collected from special selection lines [7].

Compared with the range of results published in other studies, the accuracies of GEBV for Australian Angus presented here are generally at the lower end of the range [13,14,23-25]. For example accuracies of GEBV that are commercially available from Igenity (www.igenity.com) for growth traits, carcass marbling and carcass weight in Australian/American Angus were generally higher than 0.4 [23,25]. In contrast, especially for growth traits, our values were lower than 0.1 except the accuracy of g.FW. The same applies to GEBV that are commercially available from Zoetis (www.Zoetis.com) [23,24]. In all the studies cited above, GEBV were evaluated within-breed only, but Beef CRC PE were derived across indicine and taurine breeds. Studies on beef cattle across-breed predictions are limited [14,15], but accuracies of g.CIMF and g.WW reported here were in the same range than those in [15]. However, accuracies of g.YW was ~ 0.1, whereas results of both the latter citations were between 0.3 and 0.45. Moreover, [14] found an accuracy of g.WW of 0.36, compared to our result of 0.09 from ALL PE. Differences between accuracies obtained from different PE were minor except between the BRAHMAN PE and the other three PE. The small differences in accuracies obtained from the ANGUS and BOSTAURUS PE may result from the *Bos taurus* training set consisting of almost 50% Angus individuals [7]. However, the addition of indicine breeds to the training set, which represented about 60% of the ALL PE training set, had small positive effects on the accuracies of almost all GEBV. In contrast, the BRAHMAN PE performed worst in the Angus breed for most traits, which combined to the results from ALL PE, reinforces the empirical finding that the target breed must be a member of the training population [14]. However, given the high standard errors, in general differences between accuracies obtained from different PE for a given trait were not statistically significant.

Accuracies of GEBV from ALL PE for the Limousin breed reflect that no pure-breed Limousin individuals were part of the training population. Generally, accuracies reported here do not show any consistent pattern within traits across PE. In contrast, accuracies for the American Limousin population from within-breed predictions were equal to ~0.4, and for yearling weight, they even reached 0.76 [26]. Moreover, accuracies of GEBV predicted from PE derived from a cross-breed population that consisted of only about 7% Limousin genome were between 0.2 and 0.65 depending on the trait [15].

For the Brahman breed, the only pure-breed *Bos indicus* cattle in the training population, the ALL PE yielded the highest accuracies for most GEBV, followed by the BRAHMAN PE, whereas the ANGUS and BOSTAURUS PE yielded negative results in most cases. The poor performance of the BOSTAURUS and ANGUS PE is in line with the poor performance of the BRAHMAN PE in the Angus breed, which reflects the need of having all predicted breeds in the training population. The better performance of the ALL PE compared to the BRAHMAN PE might result from additional information embedded in the LD between certain SNPs and QTL across *Bos taurus* and *Bos indicus* sub-species, in conjunction with a higher power of detection due to an increased training population size [8,22]. However, the standard errors of the accuracies do not allow for a statistically based preference of a certain PE.

## 4.2 Heritabilities

Low heritabilities of GEBV indicate that our results for the Limousin breed and partly for the Brahman breed may be affected by genotyping errors, pedigree errors or very low relationships between individuals with GEBV. For the Limousin and Brahman breeds, heritabilities varied considerably within traits across PE (e.g. for g.WW 0.5 to 0.65 for Limousin, 0.64 to 0.84 for Brahman). Since GEBV are linear functions of SNP genotypes, and SNP genotypes were the same for all PE, the heritabilities of GEBV for the same trait from different PE were expected to be equal. This assumption holds only if genotypes are obtained without errors, or if genotyping and imputation errors affect all SNPs equally. If some SNPs are more affected by errors than others and the PE weight SNPs differently, the heritabilities of GEBV for a certain trait from different PE may vary although the same animals and genotypes were used. However, this is only expected to be the case when evaluating PE in target populations because poor genotyping/imputation quality of individuals in the training population is accounted for by the prediction equation via altered GEBV accuracy. Thus, if the genotypes of validation animals were affected by imputation errors, accuracies of GEBV may increase as a result of an increased imputation accuracy.

## 4.3 Estimation of accuracies

Genomic PE are usually derived to implement GS in certain target populations by supplying the PE or GEBV to the breeding organisations. The parameter of paramount interest when evaluating PE or GEBV is the proportion of the additive genetic variance of the phenotypic target trait in the target population explained by the GEBV, where the square root of this parameter is the accuracy of the GEBV. From the perspective of the breeding organisation, this parameter can be obtained either by using the accuracy generated during the process of generating the PE, which assumes that this accuracy is equal to a variance ratio, or by re-estimating this parameter in the target population. Using the accuracy from the PE generation process bears the risk of assuming the GEBV to be more accurate than they actually are if the genetic link to the training population is insufficient [12], or if the training population sample size does not reflect the genetic variability in the target population because parameters estimated in the training population may not be valid in the target population. This problem can be circumvented by re-estimating the accuracy in the target population. However, accuracies from the process of generating the PE as well as those re-estimated in the target population can be biased due to the method of calculation. The accuracy of GEBV is often estimated as the correlation between the GEBV and a response variable, which can be breeding values, de-regressed proofs, daughter yield deviations, phenotypes or scaled versions of these variables. The co-variances necessary to calculate the correlation are obtained from inner-space vector products of the GEBV vector and the response variable vector [8,10,11]. The expectation of the inner-space vector product of two random vectors with expectation 0 is the trace of their co-variance matrix. Assuming that the co-variance matrix is a matrix times a scalar co-variance, the inner-space vector product will estimate the scalar co-variance correctly only if the average diagonal element of the matrix is 1. If the average is larger than 1, it will inflate the co-variance. Thus, if the covariance matrix between the response variable and the GEBV is the genomic relationship matrix times their covariance, and the average diagonal element of the genomic relationship matrix is larger than 1, the covariance will be biased upwards, and the accuracy will be overestimated. Moreover, genetic trends due to selection may further increase the inner-space vector product due to a mean of the random vectors larger than 0. In addition to the possible bias from transferring GEBV accuracies to the target population and from the method of calculation, the above methodology does not exploit all available phenotypic data when deriving the PE or when estimating the GEBV accuracy. PE using all available data can be derived by a single-step methodology and back-solving the single-step breeding values [27-30]. However, an accuracy in the sense of the proportion of the additive genetic variance explained by GEBV cannot be achieved from such analysis. The REML approach used in this article and by [13-16] overcomes several of the above outlined shortcomings by re-estimating the accuracy in the target population, using as much phenotypic data as available, allowing for sources of bias in parameter estimation due to relationships between individuals, selection and inbreeding, and generating the parameter of paramount interest, the proportion of the genetic variance of the target trait in the target population explained by the genetic covariance between the target trait and the GEBV.

## 4.4 Across-breed prediction

Across-breed prediction has its theoretical basis in the finding that the LD between SNPs persists over much longer genome distances within breed than across breeds, and in the assumption that trait coding QTL are the same across breeds. Thus, mixing breeds may lead to a sample with most advanced LD decay between QTL and SNPs such that across all breeds in this sample only SNPs in close proximity to the QTL are still in high LD [6,22,31]. Up to this point, this theory is supported by the fact that ALL PE worked best in Angus, followed by ANGUS, BOSTAUR and BRAHMAN PE. However, a consequence of the above logic is that the addition of a breed "N" to a training population of "N-1" breeds must be of decreasing marginal benefit, because the total probability that the LD between QTL and their adjacent SNPs is already exploited increases with every additional individual and/or breed. In practical terms, in a set of "N" breeds, GEBV for breed "N" should be predictable with high accuracy from a training set of "N-1" breeds. Violation of the equal QTL assumption does not invalidate the marginal benefit principle. It also applies when breeds have specific QTL alleles due to mutation or due to an ancestral population with more than two QTL alleles, as long as the LD phase between SNPs and negative/positive QTL alleles is the same for all breeds in the training population. Results in this paper, as well as those published from other across-breed prediction trials [8,14,32], show that all breeds in the target/validation population must be part of the training population to obtain sufficient GEBV accuracies. In the framework of the above theory and its marginal benefit consequence, the conclusion would be that the Limousin and Angus breeds are genetically more different than are the Angus and Brahman breeds, with this difference including different trait coding QTL, inversion of LD phases, fixed SNPs, and SNPs in linkage equilibrium with trait coding QTL. Since such a conclusion contradicts the phylogeny of cattle breeds, the empirical result that across-breed PE yield accurate GEBV only if all targeted breeds are in the training population does not fit into the above genetic theory.

To date, the fact that across-breed prediction works only if all targeted breeds are in the training population may result from partial or total collinearity between very distant SNP genotypes. Collinearity between SNP genotypes can be a result of a physical proximity between two

SNPs in terms of base pairs, and can persist over many generations after being induced by an ancient sampling event (e.g. the breed formation). However, collinearity may be also observed between distant SNP genotypes which can be induced by a recent sampling event, for instance sampling a number of individuals for genotyping of which SNP haplotypes do not reflect the genetic diversity in the original population, which is facilitated by the number of SNPs usually exceeding the number of genotyped animals. Such "genotype sampling collinearity" between SNP genotypes in close proximity to a QTL and SNP genotypes very distant from this location will result in LD between QTL and these very distant SNPs. When estimating SNP effects, both types of SNPs will then compete for the effect of the same QTL. Since this kind of collinearity is likely to change with every single individual or breed added to the training population, the prediction equation will change subsequently. How well prediction equations can be transferred to other populations is a function of this change. The empirical finding that targeted breeds must be members of the training population to successfully apply the prediction equation supports the conclusion that much of the LD in current across-breed data sets is induced by the sampling event which arises when individuals are chosen for genotyping.

## 5. CONCLUSIONS

Although accuracies of GEBV are generally low compared to already published accuracies that are estimated within breeds, they are in line with those derived from other across-breed prediction trials. Thus, prediction equations derived by the Beef CRC from a mixed-breed training population can contribute to the implementation of genomic selection in Australian beef cattle breeding. Since across-breed prediction equations performed equally or better than the within-breed prediction equations, and the mixed-breed dataset is likely to grow faster than the pure-breed dataset, we recommend that breeders use prediction equations from the mixed-breed training population. However, breeding organisations should only implement GS on the basis of Beef CRC across-breed equations if their breed was part of the training population.

# ACKNOWLEDGEMENTS

The authors thank Meat and Livestock Australia for supporting this work (Project B.BFG.0050). We acknowledge the Beef Cooperative Research Center, Bolormaa Sunduimijid (Victoria Department of Primary Industry, Melbourne, Australia) and Mike Goddard (University of Melbourne, Australia) for providing prediction equations. Moreover, we acknowledge Angus Australia, the Australian Brahman Breeders' Association and the Australian Limousin Breeders' Society for providing access to their data, the Irish Cattle Breeding Federation (ICBF) for providing reference genotypes for imputation and Hans Graser as the former director of AGBU for organising the collaboration with various partners and invaluable advice and discussions.


# COMPETING INTERESTS

The authors declare that they have no competing interests.

# AUTHORS' CONTRIBUTIONS

VB analysed the data, ran the variance component estimation, wrote the manuscript, responded to reviewers and editors and revised the manuscript. DJ designed the experiment and contributed to the manuscript. BT imputed the genotypes, calculated the GEBV and contributed to the manuscript. All authors read and approved the final manuscript.

# REFERENCES


1 Johnston DJ, Tier B, Graser HU: **Integration of DNA markers into BREEDPLAN EBVs**. In *Proceedings of the Association for the Advancement of Animal Breeding and Genetics 18th Conference: 28 September–1 October 2009; Barossa Valley, Australia*; 2009:30–33.

2 Hayes BJ, Bowman PJ, Chamberlain AJ, Goddard ME. **Invited review: genomic selection in dairy cattle: progress and challenges**. J Dairy Sci. 2009;92:433–443.

3 Swan AA, Johnston DJ, Brown DJ, Tier B, Graser HU. **Integration of genomic information into beef cattle and sheep genetic evaluations in Australia**. Anim Prod Sci. 2012;52:126–132.

4 Goddard ME, Hayes BJ. **Genomic selection**. J Anim Breed Genet. 2007;124:323–330.

5 Johnston DJ, Tier B, Graser HU. **Beef cattle breeding in Australia with genomics: opportunities and needs**. Anim Prod Sci. 2012;55:100–106.

6 de Roos APW, Hayes BJ, Spelman RJ, Goddard ME. **Linkage disequilibrium and persistence of phase in Holstein-Friesian, Jersey and Angus cattle**. Genetics. 2008;179:1503–1512.

7 Bolormaa S, Pryce JE, Kemper K, Savin K, Hayes BJ, Barendse W, Zhang Y, Reich CM, Mason BA, Bunch RJ, Harrison BE, Reverter A, Herd RM, Tier B, Graser HU, Goddard ME. **Accuracy of prediction of genomic breeding values for residual feed intake, carcass and meat quality traits in Bos taurus, Bos indicus and composite beef cattle**. J Anim Sci. 2013;91:3088–3104.

8 Hayes BJ, Bowman PJ, Chamberlain AC, Verbyla K, Goddard ME. **Accuracy of genomic breeding values in multi-breed dairy cattle populations**. Genet Sel Evol. 2009;41:51.

9 Saatchi M, McClure MC, McKay SD, Rolf MM, Kim J, Decker JE, Taxis TM, Chapple RH, Ramey HR, Northcutt SL, Bauck S, Woodward B, Dekkers JCM, Fernando RL, Schnabel RD, Garrick DJ, Taylor JF.**Accuracies of genomic breeding values in American Angus beef cattle using k-means clustering for cross-validation**. Genet Sel Evol. 2011;43:40.

10 Luan T, Woolliams JA, Lien S, Kent M, Svendsen M, Meuwissen THE. **The accuracy of genomic selection in Norwegian red cattle assessed by cross-validation**. Genetics. 2009;183:1119–1126.

11 Verbyla KL, Hayes BJ, Bowman PJ, Goddard ME. **Accuracy of genomic selection using stochastic search variable selection in Australian Holstein Friesian dairy cattle**. Genet Res (Camb) 2009;91:307–311.

12 Habier D, Fernando RL, Dekkers JCM. **The impact of genetic relationship information on genome-assisted breeding values**. Genetics. 2007;177:2389–2397.

13 MacNeil MD, Nkrumah JD, Woodward BW, Northcutt SL. **Genetic evaluation of Angus cattle for carcass marbling using ultrasound and genomic indicators**. J Anim Sci. 2010;88:517–522.

14 Kachman SD, Spangler ML, Bennett GL, Hanford KJ, Kuehn LA, Snelling WM, Thallman RM, Saatchi M, Garrick DJ, Schnabel RD, Taylor JF, Pollak EJ. **Comparison of molecular breeding values based on within- and across-breed training in beef cattle**. Genet Sel Evol. 2013;45:30.

15 Weber KL, Thallman RM, Keele JW, Snelling WM, Bennett GL, Smith TPL, McDaneld TG, Allan MF, Van Eenennaam AL, Kuehn LA. **Accuracy of genomic breeding values in multibreed beef cattle populations derived from deregressed breeding values and phenotypes**. J Anim Sci. 2012;90:4177–4190.

16 Weber KL, Drake DJ, Taylor JF, Garrick DJ, Kuehn LA, Thallman RM, Schnabel RD, Snelling WM, Pollak EJ, Van Eenennaam AL. **The accuracies of DNA-based estimates of genetic merit derived from Angus or multibreed beef cattle training populations**. J Anim Sci. 2012;90:4191–4202.

17 Browning BL, Browning SR. **A unified approach to genotype imputation and haplotype-phase inference for large data sets of trios and unrelated individuals**. Am J Hum Genet. 2009;84:210–223.

18 Graser HU, Tier B, Johnston DJ, Barwick SA. **Genetic evaluation for the beef industry in Australia**.Aust J Exp Agr. 2005;45:913–921.

19 R Development Core Team: **R: a Language and Environment for Statistical Computing**. Vienna: R Foundation for Statistical Computing; 2011. [http://www.R-project.org/], ISBN 3-900051-07-0.

20 Leisch F. **Sweave: dynamic generation of statistical reports using literate data analysis**. In: Härdle W, Rönz B, editors. *Compstat 2002 — Proceedings in Computational Statistics*. Heidelberg: Physica Verlag; 2002. pp. 575–580.

21 Meyer K. **Wombat: a tool for mixed model analyses in quantitative genetics by restricted maximum likelihood (REML)** J Zhejiang Univ Sci B. 2007;8:815–821.

22 de Roos APW, Hayes BJ, Goddard ME. **Reliability of genomic predictions across multiple populations**. Genetics. 2009;183:1545–1553.

23 Northcutt SL: **Genomic choices**2011,[http://www.angus.org/ AGI/ GenomicChoice11102011.pdf]

24 Johnston DJ, Jeyaruban MG, Graser HU: **Evaluation of Pfizer Animal Genetics HD 50K MVP calibration**2010,[http://agbu.une.edu.au/pdf/Pfizer_50K_September2010.pdf]

25 Börner V, Johnston DJ: **Accuracy of Igenity direct genomic values in Australian Angus**. In*Proceedings of the Association for the Advancement of Animal Breeding and Genetics 20th Conference: 20–23 October 2013; Napier, New Zealand*; 2013:211–214.

26 Saatchi M, Schnabel RD, Rolf MM, Taylor JF, Garrick DJ. **Accuracy of direct genomic breeding values for nationally evaluated traits in US Limousin and Simmental beef cattle**. Genet Sel Evol.2012;44:38.

27 Legarra A, Aguilar I, Misztal I. **A relationship matrix including full pedigree and genomic information**. J Dairy Sci. 2009; 92:4656–4663.

28 Christensen OF, Lund MS. **Genomic prediction when some animals are not genotyped**. Genet Sel Evol. 2010;42:2.

29 Strandén I, Garrick DJ. **Technical note: Derivation of equivalent computing algorithms for genomic predictions and reliabilities of animal merit**. J Dairy Sci. 2009;92:2971–2975.

30 Legarra A, Ducrocq V. **Computational strategies for national integration of phenotypic, genomic, and pedigree data in a single-step best linear unbiased prediction**. J Dairy Sci. 2012; 95:4629–4645.

31 Goddard M, Hayes B, Chamberlain HMA: **Can the same genetic markers be used in multiple breeds?** In *Proceedings of the 8th World Congress on Genetics Applied to Livestock Production: 13–18 August 2006; Belo Horizonte, Brazil*; 2006:22–16.

32 Toosi A, Fernando RL, Dekkers JCM. **Genomic selection in admixed and crossbred populations**. J Anim Sci. 2010;88:32–46.

# CHAPTER 12

# Prediction of haplotypes for ungenotyped animals and its effect on marker-assisted breeding value estimation

**Han A Mulder,** [1] **Mario PL Calus,**[1] **and Roel F Veerkamp**[1]

[1]Animal Breeding and Genomics Centre, Wageningen UR Livestock Research, PO Box 65, 8200 AB Lelystad, The Netherlands

## ABSTRACT

### Background

In livestock populations, missing genotypes on a large proportion of animals are a major problem to implement the estimation of marker-assisted breeding values using haplotypes. The objective of this article is to develop a method to predict haplotypes of animals that are not genotyped using mixed model equations and to investigate the effect of using these predicted haplotypes on the accuracy of marker-assisted breeding value estimation.

## Methods

For genotyped animals, haplotypes were determined and for each animal the number of haplotype copies (nhc) was counted, i.e. 0, 1 or 2 copies. In a mixed model framework, nhc for each haplotype were predicted for ungenotyped animals as well as for genotyped animals using the additive genetic relationship matrix. The heritability of nhc was assumed to be 0.99, allowing for minor genotyping and haplotyping errors. The predicted nhc were subsequently used in marker-assisted breeding value estimation by applying random regression on these covariables. To evaluate the method, a population was simulated with one additive QTL and an additive polygenic genetic effect. The QTL was located in the middle of a haplotype based on SNP-markers.

## Results

The accuracy of predicted haplotype copies for ungenotyped animals ranged between 0.59 and 0.64 depending on haplotype length. Because powerful BLUP-software was used, the method was computationally very efficient. The accuracy of total EBV increased for genotyped animals when marker-assisted breeding value estimation was compared with conventional breeding value estimation, but for ungenotyped animals the increase was marginal unless the heritability was smaller than 0.1. Haplotypes based on four markers yielded the highest accuracies and when only the nearest left marker was used, it yielded the lowest accuracy. The accuracy increased with increasing marker density. Accuracy of the total EBV approached that of gene-assisted BLUP when 4-marker haplotypes were used with a distance of 0.1 cM between the markers.

## Conclusions

The proposed method is computationally very efficient and suitable for marker-assisted breeding value estimation in large livestock populations including effects of a number of known QTL. Marker-assisted breeding value estimation using predicted haplotypes increases accuracy especially for traits with low heritability.

# 1. BACKGROUND

In livestock, many QTL regions have been identified for quantitative traits [1]. In some cases, fine mapping has also led to the detection of causative mutations, e.g. DGAT1 in dairy cattle for milk yield and milk composition [2,3] and IGF2 in pigs for body weight [4]. In breeding programs these QTL-regions can be utilized in marker-assisted selection (MAS). Three types of markers can be used: markers in linkage equilibrium with the QTL (LE-MAS), markers in linkage disequilibrium with the QTL (LD-MAS) and the causative mutation itself as in gene-assisted selection (GAS). GAS leads to the highest genetic gain, because no recombination exists between the marker and QTL [5]. However, identifying the gene is not easy and is resource demanding [1]. The amount of QTL variation explained by markers in LD-MAS can be increased by increasing the marker density and thereby increasing the LD between markers and QTL. Alternatively, combining alleles of different marker loci into haplotypes is expected to increase the proportion of captured QTL variance as well. Based on data of a whole genome scan with 9323 SNP-markers in Angus cattle, Hayes *et al.* [6] have reported that 4 and 6-marker haplotypes increased the accuracy of MAS more than the single marker in highest LD with the QTL. However, 2-marker haplotypes performed worse than the best marker.

One of the challenges when applying MAS in livestock populations is that often a large part of the population is not genotyped, i.e. some animals have only phenotypes, some have only genotypes and others have both genotypes and phenotypes. Several methods have been proposed to overcome these differences. For LE-MAS, one would like to apply a method that uses identity-by-descent (IBD) information of haplotypes to properly account for relationships between haplotypes of related animals and to account for phase differences between markers and QTL in different families [7]. Creation of inverse IBD-matrices is, however, very time consuming [8]. With high-density SNP-chips, LD-MAS can be applied without having to use IBD-matrices. With LD-MAS, either flanking markers or identical-by-state haplotypes (IBS) can be used in marker-assisted breeding value estimation. When using flanking markers in MAS, genotype probabilities could be calculated with iterative peeling methods [9-13] but these are time consuming. Gengler*et al.* [14,15] have proposed a straightforward and quick method to predict genotype probabilities and gene contents for bi-allelic markers using a mixed model methodology, where gene content is the number of positive (negative) alleles (i.e. 2, 1, 0 for AA, Aa, aa). For ungenotyped animals, the accuracy of predicted gene contents is similar whether mixed model equations or single-marker iterative peeling are used [8,14]. Gengler *et al.* [14] suggested that the method can also be applied in the case of multi-allelic markers. Multi-marker IBS haplotypes can be considered as a special form of multi-allelic markers, making the mixed model methodology a candidate method to predict haplotypes for ungenotyped animals.

The objective of this article is to develop a method to predict haplotypes of animals that are not genotyped using mixed model equations and to investigate the effect of using those predicted haplotypes on the accuracy of marker-assisted breeding value estimation. The method is evaluated using Monte Carlo simulation, varying haplotype length, heritability of the trait and distance between the markers. The method is compared to gene-assisted and conventional breeding value estimation, which yield, respectively, the upper and lower limit of accuracy.

# 2. METHODS

## 2.1 Prediction of haplotypes with missing genotypes

Consider a situation where a QTL-region is mapped for a trait, without having identified the causative mutation and where some animals in the population are genotyped for SNP-markers in that region, but most of them are not genotyped, which is very common in animal breeding populations. In this study we would like to use IBS-haplotypes in marker-assisted breeding value estimation. When the haplotype is based on the single SNP-marker closest to the QTL, the method of Gengler *et al.* [14,15] can be used to predict the missing 'gene content', the number of A-alleles, if there are A and a-alleles. The method of Gengler *et al.* [14,15] uses the additive genetic relationship matrix in a mixed model setting to predict the gene contents of those animals not genotyped based on genotyped relatives. This method can not be applied directly for haplotypes based on multiple markers, because discrete haplotypes can not be directly constructed based on predicted continuous gene contents of SNP-markers for ungenotyped animals. However, this procedure can be easily modified to apply to a situation with haplotypes based on multiple markers. Consider that haplotypes are based on two bi-allelic markers, one on each side of the QTL. There are four possible haplotypes. For every genotyped animal, one can infer how many copies it carries for each haplotype (*nhc* = number of haplotype copies), which is 0, 1 or 2 (see Table 1 for a small example). This is in essence the same as the 'gene content' for a bi-allelic locus and the same mixed model methodology with the additive genetic relationship matrix can be applied to predict the *nhc* for each haplotype for the ungenotyped animals. In the case of *n* haplotypes this can be modeled as:

$$nhc_i = \mu_{nhc_i} + d_i + e_{nhc_i} \qquad (1)$$

**Table 1** Example with four animals with the number of haplotype copies for two SNP-marker haplotypes

| | | | Number of haplotype copies (nhc) | | | |
|---|---|---|---|---|---|---|
| Animal | Haplotype 1 | Haplotype 2 | Hap1 (11) | Hap2 (12) | Hap3 (21) | Hap4 (22) |
| 1 | 11 | 11 | 2 | 0 | 0 | 0 |
| 2 | 11 | 12 | 1 | 1 | 0 | 0 |
| 3 | 11 | 21 | 1 | 0 | 1 | 0 |
| 4 | 11 | 22 | 1 | 0 | 0 | 1 |

where $nhc_i$ is the number of copies of haplotype *i* (which is 0, 1 or 2 effectively), $\mu_{nhc_i}$ is the population mean number of copies of haplotype *i*, $d_i$ is the EBV for $nhc_i$ and $e_{nhc_i}$ is the residual of $nhc_i$. Although $\sum_{i=1}^{n} nhc_i = 2$ for each animal, it is assumed that the haplotypes are independent from each other; therefore *n* univariate mixed model analyses can be performed. Analogous to gene contents for a bi-allelic locus [14], this can be formulated in mixed model matrix notation as:

$$\begin{bmatrix} \mathbf{1}'\mathbf{1} & \mathbf{1}'\mathbf{M} \\ \mathbf{M}'\mathbf{1} & \mathbf{M}'\mathbf{M} + \lambda\mathbf{A}^{-1} \end{bmatrix} \begin{bmatrix} \mu_{nhc} \\ \mathbf{d}_y \\ \mathbf{d}_x \end{bmatrix} = \begin{bmatrix} \mathbf{1}'\mathbf{nhc}_y \\ \mathbf{M}'\mathbf{nhc}_y \end{bmatrix} \qquad (2)$$

where **1** is a vector of ones, **M** is a design matrix linking **d** with $\mathbf{nhc}_y$, $\mathbf{A}^{-1}$ is the inverse additive genetic relationship matrix, $\lambda$ is the variance ratio of residual variance and additive genetic variance for *nhc* allowing for a small proportion of genotyping and haplotyping errors or recombination $\lambda = \sigma^2_{e_{nhc}} / \sigma^2_{a_{nhc}} = 0.01/0.99$, **d** is a vector with the EBV for *nhc* with $\mathbf{d}_y$ for genotyped animals and $\mathbf{d}_x$ for ungenotyped animals, $\mathbf{nhc}_y$ is a vector with observed *nhc* of genotyped animals and is set to missing for ungenotyped animals. The heritability assumed for *nhc* is 0.99. Basically, with no genotyping or haplotyping errors, $u_{nhc_i} + d_i$ (the predicted *nhc*) should be equal to the phenotype (the true *nhc*) for genotyped animals, implying a heritability of 1.0. In the case of haplotypes, recombinant haplotypes can be transmitted from one parent to its offspring. In such a case, the recombinant haplotype can not be fully explained in the model by the haplotypes of the parent. This decreases the parent-offspring regression, i.e. decreasing the heritability. Here we set the heritability to 0.99 to allow for some small proportions of genotyping and haplotyping errors and recombination. Preliminary analysis showed no effect when the heritability was changed to 0.95.

## 2.2 Marker-assisted breeding value estimation using predicted haplotypes

To include the effects of the haplotypes to perform marker-assisted breeding value estimation using best linear unbiased prediction (MABLUP), these *nhc* can be used as covariables in random regression, where inclusion as a random effect is preferred so that effects will be regressed towards zero when there is hardly any phenotypic information, e.g. a certain haplotype appears only in one animal with a phenotypic record. Assuming no other systematic environmental effects, the model is as follows:

$$y = \mu + u_{pol} + \sum_{i}^{n} \left( \hat{nhc}_i \times h_i \right) + e \tag{3}$$

where $y$ is the phenotype, $\mu$ is the overall mean and modeled as a fixed effect, $u_{pol}$ is the random polygenic EBV, $\hat{nhc}_i = \mu_{nhc_i} + d_i$, which is the predicted number of copies of haplotype $i$, $h_i$ is the random regression coefficient for haplotype $i$ and $e$ is the residual. In matrix notation the model can be summarized as:

$$\begin{bmatrix} \mathbf{X'X} & \mathbf{X'Z} & \mathbf{X'W} \\ \mathbf{Z'X} & \mathbf{Z'Z} + \lambda_{pol}\mathbf{A}^{-1} & \mathbf{Z'W} \\ \mathbf{W'X} & \mathbf{W'Z} & \mathbf{W'W} + \lambda_{h} \end{bmatrix} \begin{bmatrix} \mathbf{b} \\ \mathbf{u}_{pol} \\ \mathbf{h}_i \end{bmatrix} = \begin{bmatrix} \mathbf{X'y} \\ \mathbf{Z'y} \\ \mathbf{w}_i'\mathbf{y} \end{bmatrix} \tag{4}$$

where **X** and **Z** are the design matrices for fixed effects and polygenic breeding values, respectively, the matrix **W** contains the $\hat{nhc}$ for all haplotypes, $\lambda_{pol}$ and $\lambda_h$ are respectively the variance ratios for the polygenic breeding values and the random regression on $n\hat{h}c_i$, **b** is the vector with solutions for fixed effects (in this case only the mean), $\mathbf{u}_{pol}$ is the vector with $u_{pol}$ and $\mathbf{h}_i$ is the vector with $h_i$. The variance of $h_i$ is $\sigma_h^2 = 0.5\sigma_{A_{qtl}}^2$ (see Appendix for derivation), where $\sigma_{A_{qtl}}^2$ is the additive genetic QTL-variance, and the variance of $u_{pol}$ is $\sigma_{u_{pol}}^2 = \sigma_{A_{pol}}^2$, where $\sigma_{A_{pol}}^2$ is the additive genetic variance due to the polygenic effect. Equations (3) and (4) can be considered as a generalization of the method by Gengler *et al.* [14,15] to multi-allelic markers and haplotypes.

## 2.3 Evaluation of method

### 2.3.1 Simulation

Monte Carlo simulation was used to evaluate the method. The simulation scheme represented a nested full-sib half-sib design (multiple offspring per mating and dam nested within sire) with discrete generations which is common in commercial animal breeding programs. The simulation scheme was identical to that reported in Mulder *et al.* [8]. One trait was simulated with additive genetic effects of one bi-allelic QTL $A_{qtl}$, a polygenic additive genetic effect $A_{pol}$ and a residual effect $e$ ($P = A_{qtl} + A_{pol} + e$). All animals had phenotypic records. Because the method of MABLUP relies on linkage disequilibrium (LD) between markers and QTL, first, 100 generations of random mating were performed prior to the data collection scheme (generation 101 - 105).

In the first 100 generations, 50 sires and 50 dams were randomly mated each generation. The QTL and 20 bi-allelic markers were placed on one 1 M long chromosome. The QTL was placed in the middle of the chromosome and the markers were equally spaced, their distance varying from 0.1 to 5 cM. The QTL was in the middle of the marker bracket between marker 10 and 11. In the founder generation, all markers and the QTL were in linkage equilibrium and had a fixed allele frequency of 0.5. The QTL-variance $\sigma_{A_{qtl}}^2$ was set to 15% of the total genetic variance, when the allele frequency is 0.5. The allele substitution effect was set to $a = \sqrt{\sigma^2_{A_{qtl}} / 2pq}$, assuming that the allele frequencies $p$ and $q$ are 0.5, which is the case in the founder generation. Recombination rates were calculated using Haldane's mapping function [16]. During these 100 generations, some markers or the QTL became fixed due to drift.

After establishing LD, from generation 101 onwards and for each generation 50 sires and 250 dams were selected based on conventional BLUP-EBV (Equation (3) without haplotype effects) and randomly mated to produce 2,000 offspring. Each sire was mated to five dams and each dam produced four male and four female offspring, resulting in that each sire had 40 half-sib offspring, five full-sib groups of eight full-sibs. A total of five generations of phenotypic data (generation 101 - 105) were created and used in breeding value estimation (10,000 animals in total). The animals of generation 101 served as base generation in the pedigree. The generations 102 - 104 were used to create linkage disequilibrium due to selection [17].

In generation 101, simulated polygenic effects were sampled from $N(0, \sigma^2_{A_{pol}})$, where $\sigma^2_{A_{pol}}$ is the polygenic genetic variance. In subsequent generations polygenic effects were sampled from $N(0.5\, A_{pol,s} + 0.5\, A_{pol,d},\ 0.5\, \sigma^2_{A_{pol}} (1 - f_p))$, where $f_p$ is the average inbreeding coefficient of the parents. Inbreeding coefficients were calculated using the Meuwissen and Luo [18] algorithm. Residual effects were sampled from $N(0, \sigma^2_e)$, where $\sigma^2_e$ is the residual variance.

The overall heritability was set to 0.03, 0.10 or 0.30, while the QTL explained 15% of the total genetic variance when the allele frequency was 0.5 as it was in the founder generation. The phenotypic variance was 1.0 in all situations when the allele frequency of the QTL was 0.5. The realized variance of the QTL was lower due to deviations of the allele frequency from 0.5 and re-estimated in generation 101. Results were based on 200 effective replicates after discarding the replicates with minor allele frequency of the QTL in the last generation (generation 105) less than 0.05. Averaged over all effective replicates, the average allele frequency of the negative QTL-allele was 0.63 in generation 101 before selection started and deviated from 0.5, because in replicates with allele frequencies closer to 0, the QTL was more likely to become fixed in generations 101-105 due to selection. The used parameter values are listed in Table 2.

**Table 2** Parameter values for simulation

| Parameter | Default value | Alternative values |
|---|---|---|
| Number of sires per generation | 50 | |
| Number of dams per generation | 250 | |
| Total number of animals | 10,000 | |
| Number of progeny per dam | 8 | |
| Number of generations | 5 | |
| Heritability | 0.3 | 0.03 and 0.10 |
| Proportion of genetic variance explained by QTL | 0.15 | |
| Number of markers simulated | 20 | |
| Distance between markers | 0.1 cM | 0.5, 1.0, 2.0, 5.0 cM |
| Number of markers used | 10 | |
| Number of replicates | 200 | |

## 2.3.2 Haplotype methods used for marker-assisted breeding value estimation

In this study we used three types of haplotypes: 1) the closest neighboring left marker of the QTL is used as a single-marker haplotype (NM), 2) both flanking markers closest to the QTL-locus are used to form a 2-marker haplotype (HAP2) and 3) on both sides the two markers closest to the QTL are used to form a 4-marker haplotype (HAP4). In the case of NM, Equation (3) and (4) reduced to the method by Gengler *et al.* [14,15] with the difference that in this case it was not the causative mutation, but a linked marker. In addition, $\sigma_h^2 = \alpha^2$, where $\alpha$ is the allele substitution effect (see equation A1 in the Appendix), because we modeled only one SNP marker allele. The markers chosen to form haplotypes had minor allele frequencies of at least 5% in generation 105. Haplotypes were known from the simulation and thus, phasing was not needed.

## 2.3.3 Genotyping and breeding value estimation

In generation 105, the breeding program starts with MABLUP according to Equation (3) and (4) using the three different haplotype methods. We simulated three genotyping scenarios: (1) only sires and males in the last generation are genotyped and (default) (2) all males are genotyped and (3) all animals are genotyped. In scenario 1 and 2, females are not genotyped. In addition to MABLUP, gene-assisted BLUP (GABLUP) and conventional BLUP (CONBLUP) are also performed for comparison. For GABLUP, it is assumed that all animals are genotyped for the QTL. For GABLUP the model is equal to Equation (3), with the difference that the true gene content is used as *nhc* and the variance is the same as for NM. For CONBLUP, Equation (3) is used without regression on *nhc* and the variance of the additive genetic effect is set to $\sigma_u^2 = \sigma_{A_{pol}}^2 + \sigma_{A_{qtl}}^2$. For all evaluations, mixed model equations were solved using MiX99, which makes use of the preconditioned conjugate gradient algorithm [19]. The mixed model equations were considered converged when the relative difference between the left-hand and right-hand sides of the mixed model equations was smaller than $1.0 * 10^{-10}$.

Accuracies were calculated as correlations between estimated and true breeding values. The QTL-EBV was calculated as $\sum_i^n \left( \hat{nhc}_i \times h_i \right)$ for each animal. The total EBV was calculated as the sum of the QTL-EBV and the polygenic EBV. Accuracies of MABLUP were compared to those of GABLUP and CONBLUP. The accuracies of GABLUP and CONBLUP can be considered as the upper and lower limits for the MABLUP accuracy. In addition, regressions of true breeding values on estimated breeding values were calculated to get an idea of the over- (regression coefficient < 1.0) or underestimation (regression coefficient > 1.0) of the variance of EBV. Bias of estimated breeding values was calculated as estimated breeding values minus true breeding values. In addition, accuracies of $\hat{nhc}$ were calculated as correlations between estimated and true *nhc* and regressions of true on estimated *nhc* were calculated.

### 2.3.4 Proportion of QTL-variance explained by the haplotypes

The proportion of QTL-variance explained by the three different haplotypes NM, HAP2 and HAP4 was calculated to assess whether using

IBS-haplotypes was suitable. The proportion of QTL-variance explained by the haplotypes is also a measure of linkage disequilibrium between the haplotype and the QTL. For NM, the $r^2$ between the marker and the QTL can be calculated as the squared correlation between them [20]. For multi-allelic haplotypes, such as HAP2 and HAP4, $r^2$ was calculated according to Equation (2) in Hayes *et al.* [6], based on an equation for multi-allelic markers by Zhao *et al.* [21].

# 3. RESULTS

## 3.1 Analysis of haplotypes

### 3.1.1 Statistics of predicted number of haplotype copies

Table 3 shows the mean, standard deviation and mean square error (MSE) for predicted number of haplotype copies (*nhc*) for ungenotyped animals as a function of the true number of haplotype copies. For all three methods, the predicted *nhc* increased with the true *nhc* and a clear distinction was made in *nhc* between animals carrying the haplotype or not. For genotyped animals the predicted *nhc* closely resembled the true*nhc*. For ungenotyped animals, the absolute numbers decreased from NM towards HAP4, due to regression to the mean and the mean *nhc* decreased from NM towards HAP4, albeit the difference between homozygotic carrier and non-carrier is largest for HAP4. As a consequence, the MSE increased with increasing true *nhc* for HAP2 and HAP4 and for HAP4 more than for HAP2. In general, the mean *nhc*decreased with the frequency of the haplotype (results not shown).

Table 4 shows the accuracy of predicted *nhc* and the regression of true *nhc* on predicted *nhc* for ungenotyped females. The accuracy decreased from NM towards HAP4, especially for HAP4, due to recombination between genotyped ancestors and ungenotyped offspring. Especially for HAP4, the accuracy decreased when the marker distance increased, which is again due to a higher probability of recombination (results not shown). The regression of true *nhc* on predicted *nhc* was approximately 1 for NM and HAP2, but somewhat lower for HAP4, due to the lower accuracy.

**Table 3** Summary statistics of predicted number of haplotype copies for ungenotyped animals

| Haplotype method | True nhc | Mean | SD | MSE |
|---|---|---|---|---|
| NM | 0 | 0.59 | 0.08 | 0.54 |
| | 1 | 0.99 | 0.09 | 0.20 |
| | 2 | 1.43 | 0.08 | 0.52 |
| HAP2 | 0 | 0.34 | 0.06 | 0.27 |
| | 1 | 0.76 | 0.08 | 0.24 |
| | 2 | 1.24 | 0.08 | 0.75 |
| HAP4 | 0 | 0.16 | 0.04 | 0.11 |
| | 1 | 0.58 | 0.06 | 0.32 |
| | 2 | 1.13 | 0.08 | 0.90 |

Mean, standard deviation (SD) and mean square error (MSE) of predicted number of haplotype copies (*nhc*) for neighboring marker (NM), 2-marker haplotype (HAP2) and 4-marker haplotypes (HAP4) for ungenotyped animals in the last generation (females) as a function of true *nhc* (sires and males in last generation are genotyped; distance between markers is 0.1 cM, heritability is 0.30, the QTL explains 15% of the genetic variance, results are averages of 200 replicates)

**Table 4** Accuracy and regression coefficients of predicted number of haplotype copies for ungenotyped animals

| Haplotype method | Accuracy nhc (se) | | Regression1 true nhc on predicted nhc (se) | |
|---|---|---|---|---|
| NM | 0.643 | (0.003) | 1.005 | (0.004) |
| HAP2 | 0.630 | (0.007) | 0.994 | (0.022) |
| HAP4 | 0.595 | (0.012) | 0.914 | (0.038) |

Accuracy of number of haplotype copies (*nhc*) and regression of true *nhc* on predicted *nhc* for neighboring marker (NM), 2-marker haplotype (HAP2) and 4-marker haplotypes (HAP4) for ungenotyped animals in the last generation (females) (sires and males in last generation are genotyped; distance between markers is 0.1 cM, heritability is 0.30, the QTL explains 15% of the genetic variance, results are averages of 200 replicates)

[1]Regressions where the variance of the predicted *nhc* was smaller than 0.0001 were omitted (denominator of regression coefficient)

## 3.2 Proportion of QTL-variance explained by haplotype

Figure 1 shows the mean proportion of QTL variance ($r^2$) explained by the haplotype as a function of marker distance. For all three methods, $r^2$ decreased with increasing marker distance. The HAP4 method captured most of the QTL variance and NM the least. Figure 2 shows the frequency distribution of $r^2$ values for the three methods at a marker density of 0.1 cM. It shows that HAP4 had the highest proportion of replicates with $r^2$ values between 0.90 and 1.00. With NM and HAP2, a substantial proportion of replicates had $r^2$ values below 20% indicating that the haplotype explained very little QTL-variance.

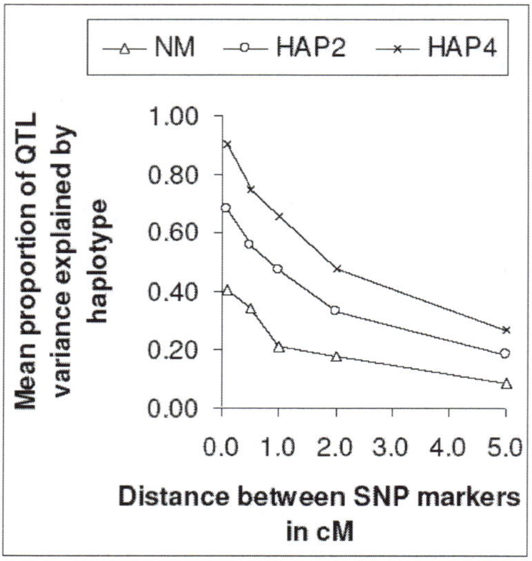


**Figure 1** Mean proportion of QTL-variance explained by haplotypes as a function of distance between SNP-markers. Mean proportion of QTL-variance explained by neighboring marker (NM), 2-marker haplotype (HAP2) and 4-marker haplotype (HAP4); average of 200 replicates.

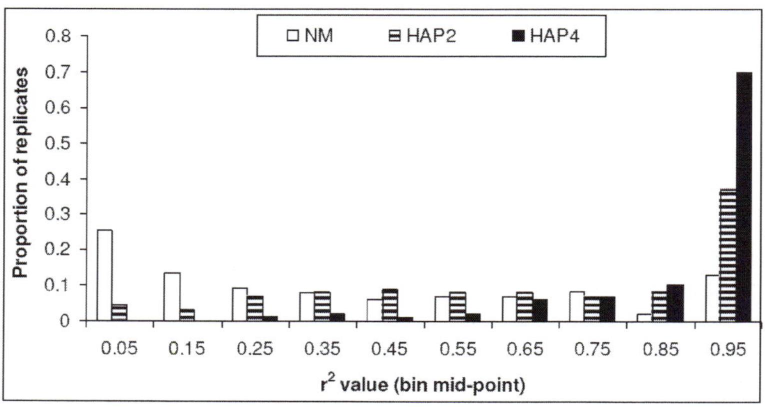


**Figure 2** Frequency distribution of QTL-variance explained by haplotypes. Proportion of replicates per 0.1-bin class of proportion of QTL variance ($r^2$) explained by neighboring marker (NM), 2-marker haplotype (HAP2) and 4-marker haplotype (HAP4); average of 200 replicates; sires and males in last generation are genotyped; distance between markers is 0.1 cM.

## 3.3 Accuracy of EBV

### 3.3.1 Effect of genotyping scenario

Table 5 shows the accuracies of QTL-EBV, polygenic EBV and total EBV for genotyped males and ungenotyped females under different genotyping scenarios with the three methods of MABLUP when the marker distance was 0.1 cM. The accuracy of polygenic and total EBV hardly changed when the number of genotyped animals increased. The accuracy of QTL-EBV increased only slightly with an increasing number of genotyped animals. This means that the use of predicted haplotypes in MABLUP did not negatively affect the accuracy of EBV. Because of the small differences in accuracy, in the rest of the article we only show results under the scenario where sires and males in the last generation were genotyped.

**Table 5** Accuracy of EBV for genotyped males and ungenotyped females in different genotyping scenarios

| | | Genotyped | | | Ungenotyped | | |
|---|---|---|---|---|---|---|---|
| EBV | Scenario2 | NM | HAP2 | HAP4 | NM | HAP2 | HAP4 |
| QTL | sires + males last | 0.534 | 0.775 | 0.912 | 0.336 | 0.491 | 0.580 |
| | all males genotyped | 0.534 | 0.774 | 0.926 | 0.337 | 0.493 | 0.591 |
| | all genotyped | 0.534 | 0.776 | 0.932 | | | |
| Polygenic | only sires + males last | 0.567 | 0.576 | 0.583 | 0.566 | 0.575 | 0.582 |
| | all males genotyped | 0.567 | 0.577 | 0.584 | 0.566 | 0.576 | 0.583 |
| | all genotyped | 0.567 | 0.578 | 0.586 | | | |
| Total | only sires + males last | 0.605 | 0.616 | 0.622 | 0.595 | 0.596 | 0.596 |
| | all males genotyped | 0.605 | 0.616 | 0.624 | 0.595 | 0.596 | 0.596 |
| | all genotyped | 0.606 | 0.617 | 0.625 | | | |

Accuracies[1] of QTL-EBV, polygenic EBV and total EBV for different genotyping scenarios for marker-assisted BLUP with neighboring marker (NM), 2-marker haplotype (HAP2) and 4-marker haplotypes (HAP4) (distance between markers is 0.1 cM, heritability is 0.30, the QTL explains 15% of the genetic variance, results are averages of 200 replicates)

[1]Standard errors were between 0.005 and 0.021 for QTL_EBV, between 0.002 and 0.003 for polygenic and total EBV; [2] in the first scenario sires from generation 101-104 and males in generation 105 were genotyped (1,200 genotyped animals); in scenario 2 all males were genotyped (5,000 genotyped animals) and in the last scenario all animals are genotyped (10,000 genotypes)

### 3.3.2 Effect of marker density

Figure 3 shows the accuracy of QTL-EBV (panel A and B) and total EBV (Panel C and D) for genotyped males (panel A and C) and ungenotyped females (panel B and D) as a function of marker distance using three different haplotype methods for MABLUP or using CONBLUP or GABLUP when all animals were genotyped. For genotyped males (Figure 3A) the accuracy of the QTL-EBV was between 0.22 and 0.90 for NM, HAP2 and HAP4 and 1.0 for GABLUP. Among the three haplotype methods, HAP4 had the highest accuracy and NM the lowest. The accuracy decreased with increasing marker distance and more rapidly for HAP4 than for NM, due to a decreasing proportion of QTL variance explained by the haplotypes (Figure 1). For ungenotyped females (Figure 3B), the accuracy of the QTL-EBV was much lower than for genotyped males, between 0.15 and 0.57 for NM, HAP2 and HAP4, but with the same trends across marker distances as for genotyped animals. The MABLUP methods based on HAP2 and HAP4 were both able to increase substantially the accuracy of the total EBV of genotyped males in comparison to CONBLUP when the distance between the markers was small (Figure 3C). The accuracy of MABLUP with HAP4 approached the accuracy of gene-assisted BLUP when the marker distance was 0.1 cM or less. The advantage of MABLUP was negligible when the marker distance was large, e.g. 5 cM. For ungenotyped animals (Figure 3D), the increase in accuracy of total EBV of MABLUP over conventional BLUP was, however, negligible regardless of marker distance.

Although the average accuracy of QTL-EBV was moderate to high for genotyped males when markers were separated by 0.1 cM, substantial variation existed between replicates (Figure 4). Especially with NM, the variation between replicates was large and even negative accuracies were obtained, although in a very small proportion of the replicates (5.5% of replicates). With HAP4, accuracies of QTL-EBV were always positive and in 86.5% of the replicates larger than 0.80. With HAP2 this proportion equaled to 60% and with NM only to 30.5%. The figure clearly shows that HAP4 had not only the highest average accuracy, but also the least variation in accuracy of QTL-EBV.

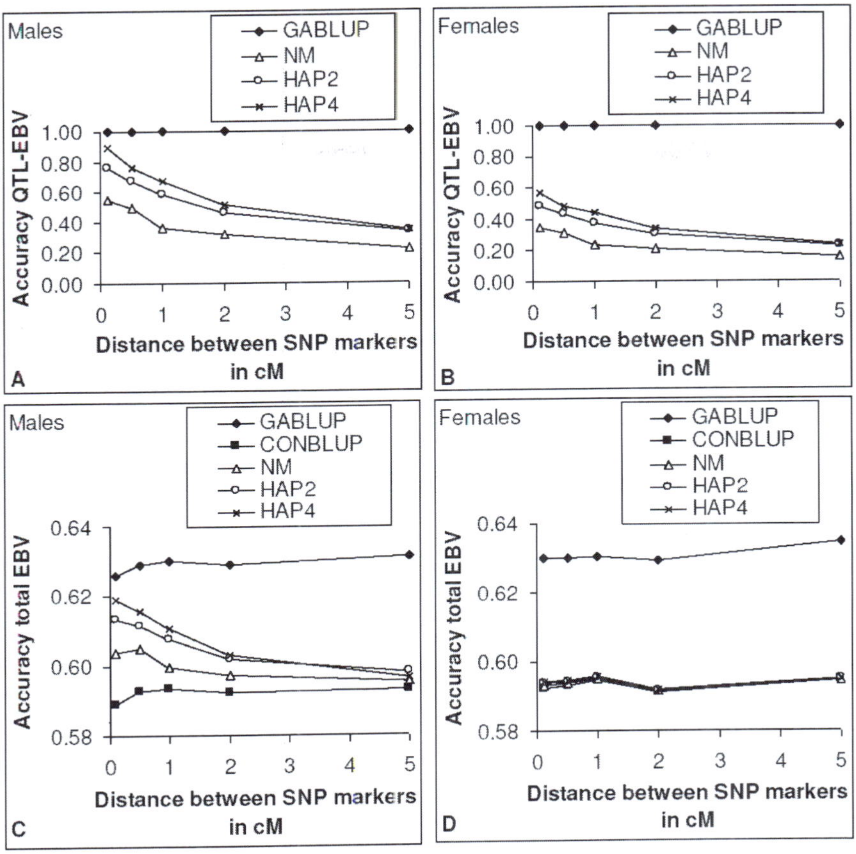


**Figure 3** Accuracy of QTL-EBV and total EBV as a function of marker distance for genotyped males and ungenotyped females. Accuracy of QTL-EBV and total EBV for marker-assisted BLUP with neighboring marker (NM), 2-marker haplotype (HAP2) and 4-marker haplotype (HAP4), geneassisted BLUP (GABLUP) when all animals are genotyped and conventional BLUP (CONBLUP); panels A and B: accuracy of QTL-EBV; panels C and D accuracy of total EBV; for MABLUP, sires and males in the last generation were genotyped, the rest was not genotyped, heritability is 0.30, the QTL explains 15% of the genetic variance, results are averages of 200 replicates.

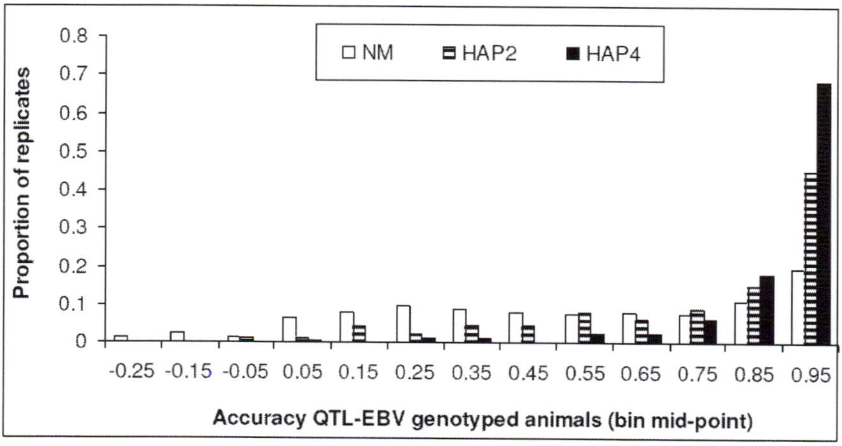


**Figure 4** Frequency distribution of accuracy of QTL-EBV of genotyped animals. Proportion of replicates per 0.1-bin-class for accuracy of QTL-EBV of genotyped animals for neighboring marker (NM), 2-marker haplotype (HAP2) and 4-marker haplotype (HAP4); sires and males in last generation are genotyped, distance between markers is 0.1 cM, heritability is 0.3, the QTL explains 15% of the genetic variance, average of 200 replicates.

### 3.3.3 Effect of heritability

Table 6 shows the accuracies of QTL-EBV, polygenic EBV and total EBV for genotyped males and ungenotyped females using different values of heritability in the three MABLUP methods when the marker distance was 0.1 cM. The accuracy of QTL-EBV increased with increasing heritability, as expected. However, the increase in accuracy of total EBV of MABLUP methods in comparison to CONBLUP was largest with a low heritability. For ungenotyped animals, the increase in accuracy with MABLUP in comparison to CONBLUP was smaller, e.g. from 0.35 to 0.37 with HAP4 at a heritability of 0.03, but the increase in accuracy was negligible when the heritability was 0.30. HAP4 had in all cases the highest accuracies for QTL-EBV, polygenic EBV and total EBV, i.e. the ranking of the methods did not change.

**Table 6** Accuracies of QTL-EBV, polygenic EBV and total EBV for genotyped males and ungenotyped females

| | | | Genotyped | | | Ungenotyped | | |
|---|---|---|---|---|---|---|---|---|
| EBV | h2 | CONBLUP | NM | HAP2 | HAP4 | NM | HAP2 | HAP4 |
| QTL | 0.03 | | 0.568 | 0.723 | 0.796 | 0.371 | 0.475 | 0.524 |
| | 0.10 | | 0.542 | 0.770 | 0.865 | 0.349 | 0.493 | 0.554 |
| | 0.30 | | 0.534 | 0.775 | 0.912 | 0.336 | 0.491 | 0.580 |
| Polygenic | 0.03 | | 0.333 | 0.336 | 0.336 | 0.335 | 0.339 | 0.339 |
| | 0.10 | | 0.444 | 0.452 | 0.456 | 0.454 | 0.444 | 0.452 |
| | 0.30 | | 0.567 | 0.576 | 0.583 | 0.566 | 0.575 | 0.582 |
| Total | 0.03 | 0.351 | 0.387 | 0.407 | 0.418 | 0.362 | 0.368 | 0.371 |
| | 0.10 | 0.465 | 0.488 | 0.508 | 0.516 | 0.468 | 0.471 | 0.472 |
| | 0.30 | 0.594 | 0.605 | 0.616 | 0.622 | 0.595 | 0.596 | 0.596 |

Accuracies[1] of QTL-EBV, polygenic EBV and total EBV for different values of heritability for marker-assisted BLUP with neighboring marker (NM), 2-marker haplotype (HAP2) and 4-marker haplotypes (HAP4) and conventional BLUP (CONBLUP) (sires and males in last generation are genotyped; distance between markers is 0.1 cM, the QTL explains 15% of the genetic variance, results are averages of 200 replicates)
[1]Standard errors were between 0.007 and 0.022 for QTL-EBV, between 0.002 and 0.006 for polygenic EBV and between 0.002 and 0.005 for total EBV

Table 7 shows the regression of true on estimated breeding values for different values of heritability for the three MABLUP methods when the marker distance was 0.1 cM for genotyped males and ungenotyped females. The regressions for QTL-EBV were substantially lower than 1.0 in the majority of the situations, except when the heritability was 0.03. This indicated that the variance of the QTL-effect was overestimated when the heritability was 0.10 and 0.30. HAP4 had regression coefficients closest to 1.0 indicating that in this case, overestimation was the smallest. Regressions for polygenic and total EBV were in most cases close to one. The variances of the polygenic EBV were slightly overestimated in all cases. The variances of the total EBV were slightly overestimated for genotyped males for CONBLUP and MABLUP and slightly underestimated for ungenotyped females with MABLUP, but overestimated with CONBLUP. Overall, the variance of total EBV was less biased with MABLUP than with CONBLUP.

Table 8 shows the bias in estimated breeding values for different values of heritability using the three MABLUP methods and CONBLUP for genotyped males and ungenotyped females when the marker distance was 0.1 cM. The polygenic EBV were on average biased upwards and the QTL-EBV were biased downwards, or in other words the QTL-effects were underestimated, but the polygenic EBV absorbed this effect. The total EBV were biased upwards for all methods when the heritability was 0.10 and 0.30, due to the shift of the estimated mean in the model, which was caused by genetic trend due to selection and the change in allele frequency of the QTL. Bias was largest for NM, whereas HAP2 and HAP4 were similar. Without selection total EBV were unbiased (results not shown). There was hardly any difference in bias between genotyped males and ungenotyped females. Adding the overall mean to the EBV removed the bias in total EBV. It can be concluded that total EBV of MABLUP and EBV of CONBLUP were biased due to selection, but this bias did no affect the ranking of animals.

**Table 7** Regression coefficients of estimated breeding values for genotyped males and ungenotyped females

| | | | Genotyped | | | Ungenotyped | | |
|---|---|---|---|---|---|---|---|---|
| EBV | h2 | CONBLUP | NM | HAP2 | HAP4 | NM | HAP2 | HAP4 |
| QTL | 0.03 | | 0.867 | 1.115 | 1.143 | 0.797 | 1.109 | 1.165 |
| | 0.10 | | 0.772 | 0.899 | 0.955 | 0.809 | 0.889 | 0.953 |
| | 0.30 | | 0.869 | 0.909 | 0.917 | 0.744 | 0.884 | 0.910 |
| Polygenic | 0.03 | | 0.945 | 0.962 | 0.970 | 0.948 | 0.965 | 0.975 |
| | 0.10 | | 0.950 | 0.973 | 0.985 | 0.951 | 0.973 | 0.985 |
| | 0.30 | | 0.951 | 0.966 | 0.976 | 0.954 | 0.965 | 0.973 |
| Total | 0.03 | 0.972 | 0.954 | 0.991 | 0.986 | 0.989 | 0.997 | 0.989 |
| | 0.10 | 0.987 | 0.981 | 0.975 | 0.974 | 1.022 | 1.014 | 1.011 |
| | 0.30 | 0.966 | 1.000 | 0.988 | 0.979 | 1.032 | 1.029 | 1.026 |

Regression[1] of true on estimated breeding values for QTL-EBV, polygenic EBV and total EBV for genotyped males and ungenotyped females for different values of heritability for marker-assisted BLUP with neighboring marker (NM), 2-marker haplotype (HAP2) and 4-marker haplotypes (HAP4) and conventional BLUP (CONBLUP) (sires and males in last generation are genotyped; distance between markers is 0.1 cM, the QTL explains 15% of the genetic variance, results are averages of 200 replicates)

[1]Standard errors were between 0.015 and 0.060 for QTL_EBV, between 0.004 and 0.015 for polygenic EBV and between 0.004 and 0.014 for total EBV; regressions where the variance of the predicted *nhc* was smaller than 0.0001 were omitted (denominator of regression coefficient)

**Table 8** Bias in estimated breeding values for genotyped males and ungenotyped females

| | | | Genotyped | | | Ungenotyped | | |
|---|---|---|---|---|---|---|---|---|
| EBV | h2 | CONBLUP | NM | HAP2 | HAP4 | NM | HAP2 | HAP4 |
| QTL | 0.03 | | -0.008 | -0.007 | -0.001 | -0.009 | -0.007 | -0.001 |
| | 0.10 | | -0.022 | -0.013 | 0.000 | -0.023 | -0.014 | -0.002 |
| | 0.30 | | -0.057 | -0.028 | 0.008 | -0.060 | -0.032 | 0.003 |
| Polygenic | 0.03 | | 0.006 | 0.002 | -0.003 | 0.006 | 0.001 | -0.003 |
| | 0.10 | | 0.064 | 0.049 | 0.035 | 0.064 | 0.049 | 0.035 |
| | 0.30 | | 0.125 | 0.086 | 0.053 | 0.126 | 0.087 | 0.055 |
| Total | 0.03 | 0.007 | -0.002 | -0.005 | -0.004 | -0.003 | -0.006 | -0.005 |
| | 0.10 | 0.042 | 0.041 | 0.035 | 0.034 | 0.041 | 0.035 | 0.034 |
| | 0.30 | 0.036 | 0.068 | 0.058 | 0.061 | 0.067 | 0.055 | 0.058 |

Bias[1] (estimated - true breeding value) in QTL-EBV, polygenic EBV and total EBV for genotyped males and ungenotyped females for different values of heritability for marker-assisted BLUP with neighboring marker (NM), 2-marker haplotype (HAP2) and 4-marker haplotypes (HAP4) and conventional BLUP (CONBLUP) (sires and males in last generation are genotyped; distance between markers is 0.1 cM, the QTL explains 15% of the genetic variance, results are averages of 200 replicates)

[1]Standard errors were between 0.003 and 0.012 for $h^2 = 0.03$, between 0.005 and 0.025 for $h^2 = 0.10$ and between 0.009 and 0.037 for $h^2 = 0.30$

# 4. DISCUSSION

In this study we developed a method to predict haplotypes of ungenotyped animals using pedigree information of genotyped animals in mixed model equations and we evaluated the use of these predicted haplotypes in marker-assisted BLUP. The method is an extension of Gengler *et al.* [14,15] to multi-allelic markers or haplotypes. The method was evaluated with Monte Carlo simulation. Clearly the predicted number of haplotype copies was regressed towards the mean and more so than the gene contents in Gengler*et al.* [14,15], especially when the frequency of a certain haplotype was low, which is more likely with longer haplotypes because of an increasing number of haplotypes. When using only a neighbor marker, the predicted gene contents were in the same range as in Gengler *et al.* [14,15]. Because of the almost-unity heritability the number of haplotype copies is hardly regressed towards the mean for genotyped animals. The accuracy of the predicted haplotypes was lower for HAP4 than for HAP2 and decreased with increasing marker distance due to the increased probability of recombination. Lowering the heritability might be an option, taking into account that the number of haplotype copies from parent to offspring is not fully heritable but subject to recombination. However, BLUP is very robust against changes in heritability and preliminary results showed no effect when the heritability was changed to 0.95.

The 4-marker haplotype gave the best results in marker-assisted breeding value estimation. It captured 90% of the QTL-variance when markers were separated by 0.1 cM. Because of this high proportion of explained QTL-variance, the proportion of QTL-variance explained by the haplotype can not increase much, and therefore we did not consider longer haplotypes. Furthermore, longer haplotypes are more subject to recombination, decreasing the accuracy of predicted number of haplotype copies. Hayes *et al.* [6] found that 6-marker haplotypes explained more QTL-variance than 4-marker haplotypes, but had much lower proportions of QTL-variance explained by the markers due to lower marker density and lower LD. Hayes *et al.* [6] found that the increase in accuracy was much higher with haplotypes than with using a neighbor marker in agreement with this study. Calus *et al.* [22] investigated the use of different definitions of haplotypes on the accuracy of genomic selection and found that with a high marker density the regression on single SNP worked almost as well as haplotypes with two markers. In their study all SNP were used for a single SNP regression, whereas in this study only one SNP was used to estimate the QTL-effect. This disfavored the neighbor marker method in our study, although the ranking of the alternatives is the same as in Calus *et al.* [22]. In the context of QTL fine-mapping, Grapes *et al.* [23] found that single marker regression with 10 markers performed worse than an IBD-method using linkage disequilibrium and linkage analysis information with a haplotype window of 10 markers, but single marker regression performed similarly when 20 markers were used. Zhao *et al.* [24] found that the power of a model with regression on two or four SNP yielded higher power to detect QTL than 2- or 4-marker haplotypes. This suggests that ranking of methods for QTL mapping might be different than for accuracy of marker-assisted or genomic selection [25].

The proportion QTL-variance explained by the haplotypes or the neighbor marker ($r^2$) was higher than in Hayes *et al.* [6]. At marker distances ranging from 0.1 to 1.0 cM, estimated $r^2$ in cattle populations have been found lower (~0.05 - 0.27) than those found in this simulation study [6,26-28]. However, in pig and poultry populations higher $r^2$ have been estimated (~0.20-0.50 in pigs and poultry) [29,30], resembling the observed$r^2$ in our study. The $r^2$ between neighbor marker and QTL or between pairs of markers followed the expected $r^2$ based on distance in cM and the effective population size [31]. The lower $r^2$ values found at short distance in cattle populations is probably due to much higher effective population sizes in the past, because LD at short distances reflects more the past effective population size [32]. As a consequence of lower LD at short distances in cattle, a higher SNP density than that used in this study is necessary to achieve in cattle the same accuracy of QTL-EBV as presented here.

Haplotypes were assumed to be unrelated in this study and it was assumed that the same QTL-allele is linked to a certain haplotype (identity-by-state = IBS). Due to recombination, linkage phases between haplotypes and QTL may be different in different families. In the context of genomic selection, Calus *et al.* [22] compared 2-marker IBS-haplotypes with 2- and 10-marker identity-by-descent haplotypes using combined linkage disequilibrium linkage analysis information (LDLA) to construct the inverse IBD-matrices. They found that IBD-haplotypes yielded higher accuracies, especially when using 10-marker windows, but at the cost of much higher computing time. The difference between IBS and IBD-haplotypes decreased with increasing marker density. Therefore, in our study it is unlikely that IBD-haplotypes would increase accuracy significantly when the distance between the markers is less than 0.1 cM.

A major disadvantage of using haplotypes is the need to phase the data. Hayes *et al.* [6] estimated the effect of haplotyping errors on the proportion of QTL-variance explained by the haplotypes in their data set and found a limited effect, but suggested that phasing errors are dependent on the data structure used. Accurate and fast algorithms are available for use in livestock populations [33,34,28]. Windig and Meuwissen [34] have shown that their algorithm is very fast and yields almost perfect haplotype reconstruction with dense marker maps in pedigreed populations. Its performance was similar to that of SIMWALK2 [35] in terms of accuracy, but with a much lower computing time. Furthermore, the presented method can accommodate haplotyping errors, e.g. by adjusting the heritability of *nhc* to a lower value, albeit at the expense of a lower accuracy.

The major advantage of the method used in this study is its computing efficiency, because optimized BLUP software can be used to predict haplotypes. The computation time was respectively ~4, 6 and 10 s for neighboring marker (NM), 2-marker haplotypes (HAP2) and 4-marker haplotypes (HAP4) to predict the genotypes/haplotypes on a dual-processor 64-bit Windows PC with 2.40 GHz and 36 GB of RAM; programs were compiled for 32-bit. Therefore, breeding companies do not need other software for imputing genotypes, which is usually much slower and much more memory intensive, prohibiting its use for large populations, e.g. with more than a million animals. An additional advantage is that no assumptions are needed on where ungenotyped animals should appear in the pedigree, it can handle all possible scenarios. Therefore, the proposed method is very suitable for application of marker-assisted breeding value estimation in large populations, such as national evaluations in cattle. Also for genomic selection purposes the method is very useful, e.g. for 50,000 SNP-markers it would take only about two days on a single processor to predict all SNP-genotypes or haplotypes for a similar number of animals as in this study.

The use of 4-marker haplotypes (HAP4) increased the accuracy of marker-assisted breeding value estimation substantially in comparison to conventional breeding value estimation for genotyped animals, but the benefit for ungenotyped animals was small in agreement with Mulder *et al.* [8]. However, with a low heritability, ungenotyped animals gained considerably in accuracy. This can be visualized by approximating the accuracy of the total EBV ($r_{totalEBV}$) as:

$$r_{totalEBV} = \sqrt{(1-q^2)r^2_{A_{pol}} + q^2 r^2_h} \qquad (5)$$

where $q^2$ is the proportion of genetic variance explained by the haplotypes $(= r^2_{A_{qtl}} \times Q^2$, where $r_{A_{qtl}}$ is the accuracy of the QTL-EBV and $Q^2$ is the amount of genetic variance explained by the QTL), $r_{A_{pol}}$ is the accuracy of the polygenic EBV and $r_h$ is the accuracy of the predicted number of haplotype copies. If we take the situation where the heritability is 0.03, the distance between markers is 0.1 cM and the QTL explains 15% of the genetic variance, $r_{A_{pol}}$ is 0.34 (Table 6) and we assume that $q^2$ is 0.10 (assuming $r_{A_{qtl}} = 0.8$ (Table 6), then Equation (5) yields $r_{totalEBV} = 0.374$, close to the value in Table 6. Using Equation (5), we can also quantify the benefit of genome-wide EBV for ungenotyped animals. Lets assume that we can explain 90% of the genetic variance by markers ($q^2 = 0.9$), then we can increase $r_{totalEBV}$ up to 0.58 assuming that $r_{A_{pol}}$ is constant. So even for ungenotyped animals genome-wide EBV can increase accuracy in comparison to conventional BLUP, especially for low heritability traits, when their paternal ancestors are genotyped.

# 5. CONCLUSIONS

In this study we show that mixed model equations can be used to predict number of haplotype copies for ungenotyped animals and these predicted number of haplotype copies can be used in marker-assisted breeding value estimation. Four-marker haplotypes give the highest accuracy for total estimated breeding values. The accuracy of the total EBV increases for genotyped animals, but for ungenotyped animals the increase is marginal unless the heritability is smaller than 0.1. The method works best when the distance between the markers is less than 1 cM. The proposed method is computationally very efficient and suitable to apply for marker-assisted breeding value estimation in large livestock populations including effects of a number of known QTL. Marker-assisted breeding value estimation using predicted haplotypes increases accuracy especially for traits with low heritability. It is expected that genomic selection for ungenotyped animals using predicted haplotypes or marker genotypes will be beneficial especially for low heritable traits.

# COMPETING INTERESTS

The authors declare that they have no competing interests.

# AUTHORS' CONTRIBUTIONS

HAM developed the method, ran the simulations and evaluations and drafted the manuscript. MPLC and RFV discussed the method and results and helped to draft the manuscript. All authors read and approved the final manuscript.

# ACKNOWLEDGEMENTS

Egbert Knol, Dieuwke Roelofs-Prins, Marc Rutten, Chris Schrooten, Addie Vereijken, Martin Lidauer, Ismo Stranden and Robin Thompson are thanked for helpful discussions about this study. We would like to thank two anonymous reviewers for giving constructive suggestions for improving the manuscript.

The work was financially supported by CRV, Hendrix Genetics, IPG and the European Commission, within the 6th Framework project SABRE, contract No. FOOD-CT-2006-016250. The text represents the authors' views and does not necessarily represent a position of the Commission who will not be liable for the use made of such information.

## REFERENCES

1 Dekkers JCM. Commercial application of marker- and gene-assisted selection in livestock: Strategies and lessons. J Anim Sci. 2004;82(E Suppl):E313–E328.

2 Grisart N, Coppieters W, Farnir F, Karim L, Ford C, Berzi P, Cambisamo N, Mni MRS, Simon P, Spelman R, Georges M, Snell R. Positional candidate cloning of a QTL in dairy cattle: identification of a missense mutation in the bovine DGAT1 gene with major effect on milk yield and composition.Genome Res. 2002;12:222–231.

3 Winter A, Kramer W, Werner FAO, Kollers S, Kata S, Durstewitz G, Buitkamp J, Womack JE, Thaller G, Fries R. Association of a lysine-232/alanine polymorphism in a bovine gene encoding acyl-CoA:diacylglycerol acyltransferase (DGAT1) with variation at a quantitative trait locus for milk fat content. Proc Nat Acad Sci USA. 2002;99:9300–9305.

4 Van Laere A-S, Nguyen M, Braunschweig M, Nezer C, Collete C, Moreau L, Archibald AL, Haley CS, Buys N, Tally M, Andersson G, Georges M, Andersson L. A regulatory mutation in IGF2 causes a major QTL effect on muscle growth in pigs. Nature. 2003;425:832–836.

5 Villanueva B, Pong-Wong R, Woolliams JA. Marker assisted selection with optimised contributions of the candidates to selection. Genet Sel Evol. 2002;34:679–703.

6 Hayes BJ, Chamberlain AJ, McPartlan H, Macleod I, Sethuraman L, Goddard ME. Accuracy of marker-assisted selection with single markers and marker haplotypes in cattle. Genet Res Camb.2007; 89:215–220.

7 Fernando RL, Grossman M. Marker assisted selection using best linear unbiased prediction. Genet Sel Evol. 1989;21:467–477.

8 Mulder HA, Meuwissen THE, Calus MPL, Veerkamp RF. The effect of missing marker genotypes on the accuracy of gene-assisted breeding value estimation: a comparison of methods. Animal. 2010;4:9–19.

9 Meuwissen THE. Determining haplotypes and IBD-probabilities from dense-marker genotypes in large complex pedigrees. Proceedings 8th World Congress Genetics Applied to Livestock Production; Belo Horizonte, Brazil. Communication. 2006. pp. 20–12.

10 Van Arendonk JAM, Smith C, Kennedy BW. Method to estimate genotype probabilities at individual loci in farm livestock. Theor Appl Genet. 1989;78:735–740.

11 Fernando RL, Stricker C, Elston RC. An efficient algorithm to compute the posterior genotypic distribution for every member of a pedigree without loops. Theor Appl Genet. 1993;87:89–93.

12 Thallman RM, Bennett GL, Keele JW, Kappes SM. Efficient computation of genotype probabilities for loci with many alleles: I. Allelic peeling. J Anim Sci. 2001;79:26–33.

13 Thallman RM, Bennett GL, Keele JW, Kappes SM. Efficient computation of genotype probabilities for loci with many alleles: II. Iterative method for large, complex pedigrees. J Anim Sci. 2001; 79:34–44.

14 Gengler N, Mayeres P, Szydlowski M. A simple method to approximate gene content in large pedigree populations: application to the myostatin gene in dual-purpose Belgian Blue cattle. Animal.2007;1:21–27.

15 Gengler N, Abras S, Verkenne C, Vanderick S, Szydlowski M, Renaville R. Accuracy of prediction of gene content in large animal populations and its use for candidate gene detection and genetic evaluation. J Dairy Sci. 2008;91:1652–1659.

16 Haldane JBS. The combination of linkage values and the calculation of distances between the loci of linked factors. J Genet. 1919;8:299–309.

17 Bulmer MG. The effect of selection on genetic variability. Am Nat. 1971;105:201–211.

18 Meuwissen THE, Luo Z. Computing inbreeding coefficients in large populations. Genet Sel Evol.1992;24:305–313.

19 Lidauer M, Stranden I. Fast and flexible program for genetic evaluation in dairy cattle. Interbull Bull.1999;20:20–25.

20 Hill WG, Robertson A. Linkage disequilibrium in finite populations. Theor Appl Genet. 1968;38:226–231.

21 Zhao H, Nettleton D, Soller M, Dekkers JCM. Evaluation of linkage disequilibrium measures between multi-allelic markers as predictors of linkage disequilibrium between markers and QTL. Genet Res Camb. 2005;86:77–87.

22 Calus MPL, Meuwissen THE, De Roos APW, Veerkamp RF. Accuracy of genomic selection using different methods to define haplotypes. Genetics. 2008;178:553–561.

23 Grapes L, Dekkers JCM, Rothschild MF, Fernando RL. Comparing linkage disequilibrium-based methods for fine mapping quantitative trait loci. Genetics. 2004;166:1561–1570.

24 Zhao HH, Fernando RL, Dekkers JCM. Power and precision of alternate methods for linkage disequilibrium mapping of quantitative trait loci. Genetics. 2007;175:1975–1986.

25 Calus MPL, Meuwissen THE, Windig JJ, Knol EF, Schrooten C, Vereijken ALJ, Veerkamp RF. Effects of the number of markers per haplotype and clustering of haplotypes on the accuracy of QTL

mapping and prediction of genomic breeding values. Genet Sel Evol. 2009;41:11.

26 De Roos APW, Hayes B, Spelman R, Goddard ME. Linkage disequilibrium and persistence of phase in Holstein-Friesian, Jersey and Angus cattle. Genetics. 2008;179:1503–1512.

27 Khatkar MS, Nicholas FW, Collins AR, Zenger KR, Cavanagh JAL, Barris W, Schnabel RD, Taylor JF, Raadsma HW. Extent of genome-wide linkage disequilibrium in Australian Holstein-Friesian cattle based on a high-density SNP panel. BMC Genomics. 2008; 9:187–194.

28 Sargolzaei M, Schenkel FS, Jansen GB, Schaeffer LR. Extent of linkage disequilibrium in Holstein cattle in North America. J Dairy Sci. 2008;91:2106–2117.

29 Du F-X, Clutter AC, Lohuis MM. Characterizing linkage disequilibirum in pig populations. Int J Biol Sci. 2007;3:166–178.

30 Andreescu C, Avendano S, Brown SR, Hassen A, Lamont SJ, Dekkers JCM. Linkage disequilibrium in related breeding lines of chickens. Genetics. 2007;177:2161–2169.

31 Sved JA. Linkage disequilibrium and homozygosity of chromosome segments in finite populations.Theor Pop Biol. 1971;2:125–141.

32 Hayes BJ, Visscher PM, McPartlan HC, Goddard ME. Novel multilocus measure of linkage disequilibrium to estimate past effective population size. Genome Res. 2003;13:635–643.

33 Scheet P, Stephens M. A fast and flexible statistical model for large-scale population genotype data: applications to inferring missing genotypes and haplotypic phase. Am J Hum Genet. 2006;78:629–644.

34 Windig JJ, Meuwissen THE. Rapid haplotype reconstruction in pedigrees with dense marker maps. J Anim Breed Genet. 2004; 121:26–39.

35 Sobel E, Lange K. Descent graphs in pedigree analysis: applications to haplotyping, location scores, and marker-sharing statistics. Am J Hum Genet. 1996;58:1323–1337.

# Index